Shantanu Das

Functional Fractional Calculus for System Identification and Controls

With 68 Figures and 11 Tables

Springer

Author
Shantanu Das
Scientist
Reactor Control Division,
BARC, Mumbai–400085
shantanu@barc.gov.in

Library of Congress Control Number: 2007934030

ISBN   978-3-540-72702-6 Springer Berlin Heidelberg New York

This work is subject to copyright. All rights are reserved, whether the whole or part of the material is concerned, specifically the rights of translation, reprinting, reuse of illustrations, recitation, broadcasting, reproduction on microfilm or in any other way, and storage in data banks. Duplication of this publication or parts thereof is permitted only under the provisions of the German Copyright Law of September 9, 1965, in its current version, and permission for use must always be obtained from Springer. Violations are liable for prosecution under the German Copyright Law.

Springer is a part of Springer Science+Business Media
springer.com
© Springer-Verlag Berlin Heidelberg 2008

The use of general descriptive names, registered names, trademarks, etc. in this publication does not imply, even in the absence of a specific statement, that such names are exempt from the relevant protective laws and regulations and therefore free for general use.

Typesetting: Integra Software Services Pvt. Ltd., India
Cover design: Erich Kirchner, Heidelberg

Printed on acid-free paper   SPIN: 12053261   89/3180/Integra   5 4 3 2 1 0

This work is dedicated to my blind father Sri Soumendra Kumar Das, my mother Purabi, my wife Nita, my son Sankalan, my sister Shantasree, brother-in-law Hemant, and to my little niece Ishita

Preface

This work is inspired by thought to have an overall fuel-efficient nuclear plant control system. I picked up the topic in 2002 while deriving the reactor control laws, which aimed at fuel efficiency. Controlling the nuclear reactor close to its natural behavior by concept of exponent shape governor, ratio control and use of logarithmic logic, aims at the fuel efficiency. The power-maneuvering trajectory is obtained by shaped-normalized-period function, and this defines the road map on which the reactor should be governed. The experience of this concept governing the Atomic Power Plant of Tarapur Atomic Power Station gives lesser overall gains compared to the older plants, where conventional proportional integral and derivative type (PID) scheme is employed. Therefore, this motivation led to design the scheme for control system than the conventional schemes to aim at overall plant efficiency. Thus, I felt the need to look beyond PID and obtained the answer in fractional order control system, requiring fractional calculus (a 300-year-old subject). This work is taken from a large number of studies on fractional calculus and here it is aimed at giving an application-oriented treatment, to understand this beautiful old new subject. The contribution in having fractional divergence concept to describe neutron flux profile in nuclear reactors and to make efficient controllers based on fractional calculus is a minor contribution in this vast (hidden) area of science. This work is aimed at to make this subject popular and acceptable to engineering and science community to appreciate the universe of wonderful mathematics, which lies between the classical integer order differentiation and integration, which till now is not much acknowledged, and is hidden from scientists and engineers.

Shantanu Das

Acknowledgments

I'm inspired by the encouragements from the director of BARC, Dr. Srikumar Banerjee, for his guidance in doing this curiosity-driven research and apply them for technological achievements. I acknowledge the encouragement received from Prof. Dr. Manoj Kumar Mitra, Dean Faculty of Engineering, and Prof. Dr. Amitava Gupta of the Jadavpur University (Power Engineering Department), for accepting the concept of fractional calculus for power plant control system, and his effort along with his PhD and ME students to develop the control system for nuclear industry, aimed at increasing the efficiency and robustness of total plant. I also acknowledge the encouragements received from Prof. Dr. Siddharta Sen, Indian Institute of Technology Kharagpur (Electrical Engineering Department) and Dr. Karabi Biswas (Jadavpur University), to make this book for ME and PhD students who will carry this knowledge for research in instrumentation and control science. From the Department of Atomic Energy, I acknowledge the encouragements received from Dr. M.S. Bhatia (LPTD BARC and HBNI Faculty, Bombay University PhD guide), Dr. Abhijit Bhattacharya (CnID), and Dr. Aulluck (CnID) for recognizing the richness and potential for research and development in physical science and control systems, especially Dr. M.S. Bhatia for his guidance to carry forward this new work for efficient nuclear power plant controls. I'm obliged to Sri A.K. Chandra, AD-R&D-ES NPCIL, for recognizing the potential of this topic and to have invited me to present the control concepts at NPCIL R&D and to have this new control scheme developed for NPCIL plants. I'm also obliged to Sri G.P. Srivatava (Director EIG-BARC) and Sri B.B. Biswas (Head of RCnD-BARC) for their guidance and encouragement in expanding the scope of this research and development with various universities and colleges, and to write this book. Lastly I thank Sri Subrata Dutta (RCS-RSD-BARC) and Dr D Datta (HPD-BARC), my batch mates of 1984 BARC Training School, to have appreciated and respected the logic in this concept of fractional calculus and to have given me immense moral support with valuable suggestions to complete this work.

Without acknowledging the work of several scientists dealing to renew and enrich this particular subject all over the globe, the work will remain incomplete,— especially Dr. Ivo, Petras Department of Informatics and process control BERG facility Technical University Kosice Slovak Republic, to have helped me to deal with doubts in digitized controller in fractional domain. I took the inspiration and learnt the subject from several presentations and works of Dr. Alain Oustaloup, CNRS-University Bordeaux, Dr. Francesco Mainardi, University Bologna Italy, Dr. Stefan G Samko University do Algarve Portugal, Dr. Katsuyuki Nishimoto, Institute of Applied Mathematics Japan, Dr. Igor Podlubny Kosice Slovak Republic, Dr. Kiran, M Kolwankar, and Dr. Anil D. Gangal, Department of Physics, University of Pune, India. The efforts of Dr. Carl F. Lorenzo, Glen Research Center Cleveland Ohio, Dr. Tom T Hartley, University of Akron, Ohio, who has popularized this old (new) subject of fractional calculus is worth acknowledging. Applied work on anomalous diffusion by Proff. R.K. Saxena (Jai Narain Vyas University Rajasthan) Dr. Santanu Saha Ray, and Proff. Dr. Rasajit Kumar Bera (Heritage Institute of Technology Kolkata), which is a source of inspiration, is also acknowledged. I have been inspired by the work on modern fractional calculus in the field of applied mathematics and applied science by Dr. M. Caputo, Dr. Rudolf Gorenflo, Dr. R. Hifler, Dr. W.G. Glockle, Dr. T.F. Nonnenmacher, Dr. R.L. Bagley, Dr. R.A. Calico, Dr. H.M. Srivastava, Dr. R.P Agarwal, Dr. P.J. Torvik, Dr. G.E.Carlson, Dr. C.A. Halijak, Dr. A. Carpinteri, Dr. K. Diethelm, Dr. A.M.A El-Sayed, Dr. Yu. Luchko, Dr. S. Rogosin, Dr. K.B. Oldham, Dr. V. Kiryakova, Dr. B. Mandelbrot, Dr. J. Spanier, Dr. Yu. N. Robotnov, Dr. K.S. Miller, Dr. B. Ross, Dr. A. Tustin, Dr. Al-Alouni, Dr. H.W. Bode, Dr. S. Manabe, Dr. S.C. Dutta Roy, Dr. W. Wyss; their work are stepping stone for applications of fractional calculus for this century. I consider these scientists as fathers of modern fractional calculus of the twenty-first century and salute them.

Shantanu Das
Scientist Reactor Control Division BARC

About the Contents of This Book

The book is organized into 10 chapters. The book aims at giving a feel of this beautiful subject of fractional calculus to scientists and engineers and should be taken as a start point for research in application of fractional calculus. The book is aimed for appreciation of this fractional calculus and thus is made as application oriented, from various science and engineering fields. Therefore, the use of too formal mathematical symbolism and mathematical formal theorem stating language are restricted. Chapters 3 and 4 give an overview of the application of fractional calculus, before dealing in detail the issues about fractional differintegrations and initialization. These two chapters deal with all types of differential operations, including fractional divergence application and usage of fractional curl. Chapter 1 is the basic introduction, dealing with development of the fractional calculus. Several definitions of fractional differintegrations and the most popular ones are introduced here; the chapter gives the feel of fractional differentiation of some functions, i.e., how they look. To aid the understanding, diagrams are given. Chapter 2 deals with the important functions relevant to fractional calculus basis. Laplace transformation is given for each function, which are important in analytical solution. Chapter 3 gives the observation of fractional calculus in physical systems (like electrical, thermal, control system, etc.) description. This chapter is made so that readers get the feel of reality. Chapter 4 is an extension of Chap. 3, where the concept of fractional divergence and curl operator is elucidated with application in nuclear reactor and electromagnetism. With this, the reader gets a broad feeling about the subject's wide applicability in the field of science and engineering. Chapter 5 is dedicated to insight of fractional integration fractional differentiation and fractional differintegral with physical and geometric meaning for these processes. In this chapter, the concept of generating function is presented, which gives the transfer function realization for digital realization in real time application of controls. Chapter 6 tries to generalize the concept of initialization function, which actually embeds hereditary and history of the function. Here, attempt is made to give some light to decomposition properties of the fractional differintegration. Generalization is called as the fractional calculus theory, with the initialization function which becomes the general theory and does cover the integer order classical calculus. In this chapter, the fundamental fractional differential equation is taken and the impulse response to that is obtained. Chapter 7 gives the Laplace transform theory—a general treatment to cover initialization aspects. In this chapter, the concept of w-plane on which the fractional control system properties are studied is described. In Chap. 7 elaborate dealing is carried on for scalar initialization and vector initialization problems. In Chaps. 6 and 7 elaborate block diagrams are given to aid the understanding of these concepts. Chapter 8 gives the application of fractional calculus in electrical circuits and electronic circuits. Chapter 9 deals with the application of fractional calculus in other fields of science and engineering for system modeling and control. In this chapter, the modern aspects of multivariate controls are touched to show the applicability in fractional feedback controllers and state observer issues. Chapter 10 gives a detailed treatment of the order of a system and its identification approach, with concepts of fractional resonance, and ultra-damped and hyper-damped systems. Also a brief is presented on future formalization of research and development for variable order differintegrations and continuous order controller that generalizes conventional control system. Bibliography gives list of important and few recent publications, of several works on this old (new) subject. It is not possible to include all the work done on this subject since past 300 years. Undoubtedly, this is an emerging area or research (not so popular at present in India), but the next decade will see the plethora of applications based on this field. May be the twenty-first century will speak the language of nature, that is, fractional calculus.

Shantanu Das

Contents

1 Introduction to Fractional Calculus 1
  1.1 Introduction 1
  1.2 Birth of Fractional Calculus 1
  1.3 Fractional Calculus a Generalization of Integer Order Calculus 2
  1.4 Historical Development of Fractional Calculus 3
     1.4.1 The Popular Definitions of Fractional Derivatives/Integrals in Fractional Calculus 7
  1.5 About Fractional Integration Derivatives and Differintegration 9
     1.5.1 Fractional Integration Riemann–Liouville (RL) 9
     1.5.2 Fractional Derivatives Riemann–Liouville (RL) Left Hand Definition (LHD) 10
     1.5.3 Fractional Derivatives Caputo Right Hand Definition (RHD) .. 10
     1.5.4 Fractional Differintegrals Grunwald Letnikov (GL) 12
     1.5.5 Composition and Property 14
     1.5.6 Fractional Derivative for Some Standard Function 15
  1.6 Solution of Fractional Differential Equations 16
  1.7 A Thought Experiment 16
  1.8 Quotable Quotes About Fractional Calculus 17
  1.9 Concluding Comments 18

2 Functions Used in Fractional Calculus 19
  2.1 Introduction 19
  2.2 Functions for the Fractional Calculus 19
     2.2.1 Gamma Function 19
     2.2.2 Mittag-Leffler Function 22
     2.2.3 Agarwal Function 27
     2.2.4 Erdelyi's Function 27
     2.2.5 Robotnov–Hartley Function 27
     2.2.6 Miller–Ross Function 27
     2.2.7 Generalized R Function and G Function 28
  2.3 List of Laplace and Inverse Laplace Transforms Related to Fractional Calculus 30
  2.4 Concluding Comments 33

3 Observation of Fractional Calculus in Physical System Description ... 35
- 3.1 Introduction ... 35
- 3.2 Temperature–Heat Flux Relationship for Heat Flowing in Semi-infinite Conductor ... 35
- 3.3 Single Thermocouple Junction Temperature in Measurement of Heat Flux ... 38
- 3.4 Heat Transfer ... 40
- 3.5 Driving Point Impedance of Semi-infinite Lossy Transmission Line ... 43
  - 3.5.1 Practical Application of the Semi-infinite Line in Circuits ... 49
  - 3.5.2 Application of Fractional Integral and Fractional Differentiator Circuit in Control System ... 52
- 3.6 Semi-infinite Lossless Transmission Line ... 54
- 3.7 The Concept of System Order and Initialization Function ... 60
- 3.8 Concluding Comments ... 61

4 Concept of Fractional Divergence and Fractional Curl ... 63
- 4.1 Introduction ... 63
- 4.2 Concept Of Fractional Divergence for Particle Flux ... 63
- 4.3 Fractional Kinetic Equation ... 65
- 4.4 Nuclear Reactor Neutron Flux Description ... 67
- 4.5 Classical Constitutive Neutron Diffusion Equation ... 67
  - 4.5.1 Discussion on Classical Constitutive Equations ... 68
  - 4.5.2 Graphical Explanation ... 69
  - 4.5.3 About Surface Flux Curvature ... 69
  - 4.5.4 Statistical and Geometrical Explanation for Non-local Divergence ... 70
- 4.6 Fractional Divergence in Neutron Diffusion Equations ... 71
  - 4.6.1 Solution of Classical Constitutive Neutron Diffusion Equation (Integer Order) ... 73
  - 4.6.2 Solution of Fractional Divergence Based Neutron Diffusion Equation (Fractional Order) ... 74
  - 4.6.3 Fractional Geometrical Buckling and Non-point Reactor Kinetics ... 76
- 4.7 Concept of Fractional Curl in Electromagnetics ... 76
  - 4.7.1 Duality of Solutions ... 77
  - 4.7.2 Fractional Curl Operator ... 77
  - 4.7.3 Wave Propagation in Unbounded Chiral Medium ... 77
- 4.8 Concluding Comments ... 79

5 Fractional Differintegrations: Insight Concepts ... 81
- 5.1 Introduction ... 81
- 5.2 Symbol Standardization and Description for Differintegration ... 81
- 5.3 Reimann–Liouville Fractional Differintegral ... 82
  - 5.3.1 Scale Transformation ... 82
  - 5.3.2 Convolution ... 85

		5.3.3	Practical Example of RL Differintegration in Electrical Circuit Element Description 87
5.4			Grunwald–Letnikov Fractional Differinteration 90
5.5			Unification of Differintegration Through Binomial Coefficients 92
5.6			Short Memory Principle: A Moving Start Point Approximation and Its Error 95
5.7			Matrix Approach to Discretize Fractional Differintegration and Weights 97
5.8			Infinitesimal Element Geometrical Interpretation of Fractional Differintegrations 98
		5.8.1	Integration 99
		5.8.2	Differentiation 100
5.9			Advance Digital Algorithms Realization for Fractional Controls 102
		5.9.1	Concept of Generating Function 102
		5.9.2	Digital Filter Realization by Rational Function Approximation for Fractional Operator 103
		5.9.3	Filter Stability Consideration 106
5.10			Local Fractional Derivatives 106
5.11			Concluding Comments 107

6 Initialized Differintegrals and Generalized Calculus 109
6.1		Introduction 109
6.2		Notations of Differintegrals 110
6.3		Requirement of Initialization 110
6.4		Initialization Fractional Integration (Riemann–Liouville Approach) 112
	6.4.1	Terminal Initialization 113
	6.4.2	Side Initialization 114
6.5		Initializing Fractional Derivative (Riemann–Liouvelle Approach) 115
	6.5.1	Terminal Initialization 116
	6.5.2	Side Initialization 117
6.6		Initializing Fractional Differintegrals (Grunwald–Letnikov Approach) . 118
6.7		Properties and Criteria for Generalized Differintegrals 119
	6.7.1	Terminal Charging 121
	6.7.2	Side Charging 122
6.8		The Fundamental Fractional Order Differential Equation 122
	6.8.1	The Generalized Impulse Response Function 123
6.9		Concluding Comments 127

7 Generalized Laplace Transform for Fractional Differintegrals 129
7.1		Introduction 129
7.2		Recalling Laplace Transform Fundamentals 129
7.3		Laplace Transform of Fractional Integrals 131
	7.3.1	Decomposition of Fractional Integral in Integer Order 132
	7.3.2	Decomposition of Fractional Order Integral in Fractional Order 135

7.4		Laplace Transformation of Fractional Derivatives 136	
	7.4.1	Decomposition of Fractional Order Derivative in Integer Order	138
	7.4.2	Decomposition of Fractional Derivative in Fractional Order ...	141
	7.4.3	Effect of Terminal Charging on Laplace Transforms	142
7.5		Start Point Shift Effect 143	
	7.5.1	Fractional Integral 143	
	7.5.2	Fractional Derivative 143	
7.6		Laplace Transform of Initialization Function 144	
	7.6.1	Fractional Integral 144	
	7.6.2	Fractional Derivative 144	
7.7		Examples of Initialization in Fractional Differential Equations........ 144	
7.8		Problem of Scalar Initialization.................................. 147	
7.9		Problem of Vector Initialization 149	
7.10		Laplace Transform $s \to w$ Plane for Fractional Controls Stability..... 151	
7.11		Rational Approximations of Fractional Laplace Operator 153	
7.12		Concluding Comments.. 155	

8 Application of Generalized Fractional Calculus in Electrical Circuit Analysis 157

8.1	Introduction 157	
8.2	Electronics Operational Amplifier Circuits 157	
	8.2.1	Operational Amplifier Circuit with Lumped Components 157
	8.2.2	Operational Amplifier Integrator with Lumped Element....... 158
	8.2.3	Operational Amplifier Integrator with Distributed Element 159
	8.2.4	Operational Amplifier Differential Circuit with Lumped Elements 161
	8.2.5	Operational Amplifier Differentiator with Distributed Element . 162
	8.2.6	Operational Amplifier as Zero-Order Gain with Lumped Components 163
	8.2.7	Operational Amplifier as Zero-Order Gain with Distributed Elements 163
	8.2.8	Operational Amplifier Circuit for Semi-differintegration by Semi-infinite Lossy Line 164
	8.2.9	Operational Amplifier Circuit for Semi-integrator 165
	8.2.10	Operational Amplifier Circuit for Semi-differentiator 166
	8.2.11	Cascaded Semi-integrators 167
	8.2.12	Semi-integrator Series with Semi-differentiator Circuit 167
8.3	Battery Dynamics 168	
	8.3.1	Battery as Fractional Order System 168
	8.3.2	Battery Charging Phase................................... 168
	8.3.3	Battery Discharge Phase 172
8.4	Tracking Filter 174	
	8.4.1	Observations... 176
8.5	Fractional Order State Vector Representation in Circuit Theory 177	
8.6	Concluding Comments.. 180	

9 Application of Generalized Fractional Calculus in Other Science and Engineering Fields ... 181
9.1 Introduction ... 181
9.2 Diffusion Model in Electrochemistry ... 181
9.3 Electrode–Electrolyte Interface Impedance ... 182
9.4 Capacitor Theory ... 184
9.5 Fractance Circuit ... 185
9.6 Feedback Control System ... 187
    9.6.1 Concept of Iso-damping ... 194
    9.6.2 Fractional Vector Feedback Controller ... 196
    9.6.3 Observer in Fractional Vector System ... 197
    9.6.4 Modern Aspects of Fractional Control ... 199
9.7 Viscoelasticity (Stress–Strain) ... 200
9.8 Vibration Damping System ... 202
9.9 Concluding Comments ... 204

10 System Order Identification and Control ... 205
10.1 Introduction ... 205
10.2 Fractional Order Systems ... 205
10.3 Continuous Order Distribution ... 207
10.4 Determination of Order Distribution from Frequency Domain Experimental Data ... 209
10.5 Analysis of Continuous Order Distribution ... 211
10.6 Variable Order System ... 220
    10.6.1 RL Definition for Variable Order ... 220
    10.6.2 Laplace Transforms and Transfer Function of Variable Order System ... 222
    10.6.3 GL Definition for Variable Order ... 223
10.7 Generalized PID Controls ... 224
10.8 Continuum Order Feedback Control System ... 226
10.9 Time Domain Response of Sinusoidal Inputs for Fractional Order Operator ... 228
10.10 Frequency Domain Response of Sinusoidal Inputs for Fractional Order Operator ... 229
10.11 Ultra-damped System Response ... 229
10.12 Hyper-damped System Response ... 230
10.13 Disadvantage of Fractional Order System ... 231
10.14 Concluding Comments ... 232

Bibliography ... 233

Chapter 1
Introduction to Fractional Calculus

1.1 Introduction

Fractional calculus is three centuries old as the conventional calculus, but not very popular among science and/or engineering community. The beauty of this subject is that fractional derivatives (and integrals) are not a local (or point) property (or quantity). Thereby this considers the history and non-local distributed effects. In other words, perhaps this subject translates the reality of nature better! Therefore to make this subject available as popular subject to science and engineering community, it adds another dimension to understand or describe basic nature in a better way. Perhaps fractional calculus is what nature understands, and to talk with nature in this language is therefore efficient. For past three centuries, this subject was with mathematicians, and only in last few years, this was pulled to several (applied) fields of engineering and science and economics. However, recent attempt is on to have the definition of fractional derivative as local operator specifically to fractal science theory. Next decade will see several applications based on this 300 years (old) new subject, which can be thought of as superset of fractional differintegral calculus, the conventional integer order calculus being a part of it. Differintegration is an operator doing differentiation and sometimes integrations, in a general sense. In this book, fractional order is limited to only real numbers; the complex order differentigrations are not touched. Also the applications and discussions are limited to fixed fractional order differintegrals, and the variable order of differintegration is kept as a future research subject. Perhaps the fractional calculus will be the calculus of twenty-first century. In this book, attempt is made to make this topic application oriented for regular science and engineering applications. Therefore, rigorous mathematics is kept minimal. In this introductory chapter, list in tabular form is provided to readers to have feel of the fractional derivatives of some commonly occurring functions.

1.2 Birth of Fractional Calculus

In a letter dated 30th September 1695, L'Hopital wrote to Leibniz asking him a particular notation that he had used in his publication for the nth derivative of a function

$$\frac{D^n f(x)}{Dx^n}$$

i.e., what would the result be if $n = 1/2$. Leibniz's response " an apparent paradox from which one day useful consequences will be drawn." In these words, fractional calculus was born. Studies over the intervening 300 years have proved at least half right. It is clear that within the twentieth century especially numerous applications have been found. However, these applications and mathematical background surrounding fractional calculus are far from paradoxical. While the physical meaning is difficult to grasp, the definitions are no more rigorous than integer order counterpart.

1.3 Fractional Calculus a Generalization of Integer Order Calculus

Let us consider n an integer and when we say x^n we quickly visualize x multiply n times will give the result. Now we still get a result if n is not an integer but fail to visualize how. Like to visualize 2^π is hard to visualize, but it exists. Similarly the fractional derivative we may say now as

$$\frac{d^\pi}{dx^\pi} f(x)$$

though hard to visualize (presently), does exist. As real numbers exist between the integers so does fractional differintegrals do exist between conventional integer order derivatives and n-fold integrations. We see the following generalization from integer to real number on number line as

$$x^n = \underbrace{x.x.x.x \ldots \ldots \ldots x}_{n} \quad n \text{ is integer}$$

$$x^n = e^{n \ln x} \quad n \text{ is real number}$$

$$n! = 1.2.3 \ldots (n-1)n \quad n \text{ is integer}$$

$$n! = \Gamma(n+1) \quad n \text{ is real} \quad \text{and Gamma Functional is } \Gamma(x) = \int_0^\infty e^{-t} t^{x-1} dt$$

Therefore, the above generalization from integer to non-integer is what is making number line general (i.e., not restricting to only integers). Figure. 1.1 demonstrates the number line and the extension of this to map any fractional differintegrals. The negative side extends to say integration and positive side to differentiation.

$$f, \frac{df}{dt}, \frac{d^2 f}{dt^2}, \frac{d^3 f}{dt^3}, \ldots \rightarrow$$

1.4 Historical Development of Fractional Calculus

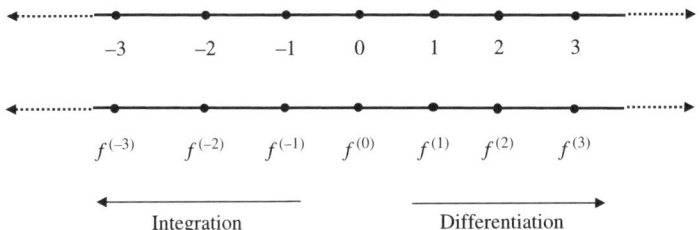

Fig. 1.1 Number line and Interpolation of the same to differintegrals of fractional calculus

$$\leftarrow \ldots, \int dt \int dt \int f\,dt, \int dt \int f\,dt, \int f\,dt, f$$

Writing the same in differintegral notation as represented in number line we have

$$\leftarrow \ldots \frac{d^{-3}f}{dt^{-3}}, \frac{d^{-2}f}{dt^{-2}}, \frac{d^{-1}f}{dt^{-1}}, f, \frac{df}{dt}, \frac{d^2 f}{dt^2}, \frac{d^3 f}{dt^3}, \ldots \rightarrow$$

Heaviside (1871) states that there is a universe of mathematics lying between the complete differentiation and integration, and that fractional operators push themselves forward sometimes and are just as real as others.

Mathematics is an art of giving things misleading names. The beautiful—and at first glance mysterious—name, the fractional calculus is just one of those misnomers, which are the essence of mathematics. We know such names as natural numbers and real numbers. We use them very often; let us think for a moment about these names. The notion of natural number is a natural abstraction, but it is the number natural itself a natural? The notion of a real number is generalization of the notion of a natural number. The real number emphasizes that we pretend that they reflect real quantities, but cannot change the fact that they do not exist. If one wants to compute something, then one immediately discovers that there is no place for real numbers in this real world. On a computer he/she can work with finite set of finite fractions, which serves as approximations to unreal real number.

Fractional calculus does not mean the calculus of fractions, nor does it mean a fraction of any calculus differentiation, integration, or calculus of variations. The fractional calculus is a name of theory of integrations and derivatives of arbitrary order, which unify and generalize the notion of integer order differentiation and n-fold integration. So we call it generalized differintegrals.

1.4 Historical Development of Fractional Calculus

Fractional order systems, or systems containing fractional derivatives and integrals, have been studied by many in engineering and science area—Heaviside (1922), Bush (1929), Goldman (1949), Holbrook (1966), Starkey (1954), Carslaw and Jeager (1948), Scott (1955), and Mikuniski (1959). Oldham and Spanier (1974) and Miller and Ross (1993) present additionally very reliable discussions devoted specifically to the subject. It should be noted that there are growing number of physical systems whose behavior can be compactly described using fractional calculus system theory. Of specific interest to electrical engineers are long electrical lines (Heaviside 1922), electrochemical process (Ichise, Nagayanagi and Kojima 1971; Sun, Onaral, and Tsao 1984), dielectric polarization (Sun, Abdelwahab and Onaral 1984), colored noise (Manderbolt 1967), viscoelastic materials (Bagley and Calico 1991; Koeller 1986; Skaar, Michel, and Miller 1988) and chaos (Hartley, Lorenzo and Qammer 1995), and electromagnetism fractional poles (Engheta 1998). During the development of the fractional calculus applied theory, for past 300 years, the contributions from N.Ya. Sonnin (1869), A.V. Letnikov (1872), H. Laurent (1884), N. Nekrasov (1888), K. Nishimoto (1987), Srivastava (1968, 1994), R.P. Agarwal (1953), S.C. Dutta Roy (1967), Miller and Ross (1993), Kolwankar and Gangal (1994), Oustaloup (1994), L.Debnath (1992), Igor Podlubny (2003), Carl Lorenzo (1998) Tom Hartley (1998), R.K. Saxena (2002), Mainardi (1991), S. Saha Ray and R.K. Bera (2005), and several others are notable. The author has tried to apply the fractional calculus concepts to describe the nuclear reactor constitutive laws and apply the theory for obtaining efficient automatic control for nuclear power plants. Following are some of the notations and formalization efforts by several mathematicians, since late seventeenth century:

Since 1695, after L'Hopital's question regarding the order of the differentiation, Leibniz was the first to start in this direction. Leibniz (1695–1697) mentioned a possible approach to fractional order differentiation, in a sense that for non-integer (n) the definition could be following. He wrote this letter to J. Wallis and J. Bernulli.

$$\frac{d^n e^{mx}}{dx^n} = m^n e^{mx}$$

L. Euler (1730) suggested using a relationship for negative or non-integer (rational) values; taking $m = 1$ and $n = 1/2$, he obtained the following:

$$\frac{d^n x^m}{dx^n} = m(m-1)(m-2)\ldots(m-n+1)x^{m-n}$$
$$\Gamma(m+1) = m(m-1)\ldots(m-n+1)\Gamma(m-n+1)$$
$$\frac{d^n x^m}{dx^n} = \frac{\Gamma(m+1)}{\Gamma(m-n+1)} x^{m-n}$$
$$\frac{d^{1/2} x}{dx^{1/2}} = \sqrt{\frac{4x}{\pi}} = \frac{2}{\sqrt{\pi}} x^{1/2}$$

First step in generalization of notation for differentiation of arbitrary function was conceived by J.B.J. Fourier (1820–1822), after the introduction of

1.4 Historical Development of Fractional Calculus

$$f(x) = \frac{1}{2\pi} \int_{-\infty}^{+\infty} f(z)dz \int_{-\infty}^{+\infty} \cos(px - pz)dp .$$

He made a remark as

$$\frac{d^n f(x)}{dx^n} = \frac{1}{2\pi} \int_{-\infty}^{+\infty} f(z)dz \int_{-\infty}^{+\infty} \cos(px - pz + n\frac{\pi}{2})dp,$$

and this relationship could serve as a definition of nth order derivative for non-integer order n. N.H. Abel (1823–1826) introduced the integral as

$$\int_0^x \frac{S'(\eta)d\eta}{(x-\eta)^\alpha} = \psi(x).$$

He in fact solved the integral for an arbitrary α and not just for $1/2$, and he obtained

$$S(x) = \frac{\sin(\pi\alpha)}{\pi} x^\alpha \int_0^1 \frac{\psi(xt)}{(1-t)^{1-\alpha}} dt.$$

After that, Abel expressed the obtained solution with the help of an integral of order of α.

$$S(x) = \frac{1}{\Gamma(1-\alpha)} \frac{d^{-\alpha}\psi(x)}{dx^{-\alpha}}.$$

J. Liouvilli (1832–1855) gave three approaches. The first one is Leibniz's formulation, which is as follows.

$$\frac{d^m e^{ax}}{dx^n} = a^m e^{ax}$$

$$f(x) = \sum_{n=0}^{\infty} c_n e^{a_n x}$$

$$\frac{d^\gamma f(x)}{dx^\gamma} = \sum_{n=0}^{\infty} c_n a_n^\gamma e^{a_n x}$$

Here, the function is decomposed by infinite set of exponential functions. J. Liouville introduced the integral of non-integer order as the second approach, which is noted below:

$$\int^{\mu} \phi(x) dx^{\mu} = \frac{1}{(-1)^{\mu}\Gamma(\mu)} \int_{0}^{\infty} \phi(x+\alpha)\alpha^{\mu-1} d\alpha$$

$$\int^{\mu} \phi(x) dx^{\mu} = \frac{1}{\Gamma(\mu)} \int_{0}^{\infty} \phi(x-\alpha)\alpha^{\mu-1} d\alpha$$

$$\tau = x + \alpha, \, \& \tau = x - \alpha$$

$$\int^{\mu} \phi(x) dx^{\mu} = \frac{1}{(-1)^{\mu}\Gamma(\mu)} \int_{x}^{\infty} (\tau-x)^{\mu-1} \phi(\tau) d\tau$$

$$\int^{\mu} \phi(x) dx^{\mu} = \frac{1}{\Gamma(\mu)} \int_{-\infty}^{x} (x-\tau)^{\mu-1} \phi(\tau) d\tau$$

The third approach given by Liouville is the definitions of derivatives of non-integer order as

$$\frac{d^{\mu} F(x)}{dx^{\mu}} = \frac{(-1)^{\mu}}{h^{\mu}} \left(F(x) - \frac{\mu}{1} F(x+h) + \frac{\mu(\mu-1)}{1.2} F(x+2h) - \ldots \right)$$

$$\frac{d^{\mu} F(x)}{dx^{\mu}} = \frac{1}{h^{\mu}} \left(F(x) - \frac{\mu}{1} F(x-h) + \frac{\mu(\mu-1)}{1.2} F(x-2h) - \ldots \right)$$

$\lim . h \to 0$

Liouville was the first to point the existence of the right-sided and left-sided differentials and integrals.

G.F.B. Riemann (1847) used a generalization of Taylor series for obtaining a formula for fractional order integration. Riemann introduced an arbitrary "complimentary" function $\psi(x)$ because he did not fix the lower bound of integration. He could not solve this disadvantage. From here, the initialized fractional calculus was born lately in the later half of the twentieth century; Riemann's notation is as follows with the complimentary function.

$$D^{-\nu} f(x) = \frac{1}{\Gamma(\nu)} \int_{c}^{x} (x-t)^{\nu-1} f(t) dt + \psi(t)$$

Cauchy formula for nth derivative in complex variables is

$$f^n(z) = \frac{n!}{j2\pi} \oint \frac{f(t)}{(t-z)^{n+1}} dt$$

and for non-integer $n = \upsilon$, a branch point of the function $(t-z)^{-\upsilon-1}$ appears instead of pole

1.4 Historical Development of Fractional Calculus

$$D^\nu f(z) = \frac{\Gamma(\nu+1)}{j2\pi} \int_C^{x+} \frac{f(t)}{(t-z)^{\nu+1}} dt$$

Generally, to understand the dynamics of any particular system, we often consider the nature of the complex domain singularities (poles). Consider a complex function $G(z) = (z^q + a)^{-1}$, where $q > 0$ and is a fractional number. This particular function of the complex variable does not have any pole on the primary Riemann sheet of the complex plane $z = r\exp(j\theta)$, i.e., within $|\theta| < \pi$. It is impossible to force the denominator $z^q + a$ to zero anywhere in complex plane $|\theta| < \pi$. Consider for $q = 0.5$, the denominator $z^{0.5} + 1$ does not go anywhere to zero in the primary Riemann sheet, $|\theta| < \pi$. It becomes zero on secondary Riemann sheet at $z = \exp(\pm j2\pi) = 1 + j0$.

Normally, to get to the secondary Riemann sheet, it is necessary to go through a "branch-cut" on the primary Riemann sheet. This is accomplished by increasing the angle in the complex plane z. Increasing the angle to $\theta = +\pi$ gets us to the "branch-cut" on the z – complex plane. This can also be accomplished by decreasing the angle until $\theta = -\pi$, which also gets us to the "branch-cut". This "branch-cut" lies at $z = r\exp(\pm j\pi)$, for all positive r. Increasing the angle further eventually gets to $\theta = \pm j2\pi$. Further increasing the angle $\theta > \pi$ makes to go "underneath" the primary Riemann sheet, inside the negative real axis of z – complex plane.

The behavior of the function $(z^{0.5}+1)^{-1}$ is thus described by two Riemann sheets. Returning to the first Riemann sheet on the z – complex plane, the branch cut begins at $z = 0$, the origin, and extends out to the negative real axis to infinity. The end of the branch cuts are called "branch-points," which are then at the origin and at minus infinity in the z – plane.

The "branch-points" can be considered as singularities on the primary Riemann sheet of the z – plane as well, but the function $(z^{0.5}+1)^{-1}$ does not go to infinity then. Therefore to obtain the plot of the pole, one has to wrap around these branch-points and go to secondary Riemann sheet (in this case, at $1 + j0$ at $\theta = \pm 2\pi$).

1.4.1 The Popular Definitions of Fractional Derivatives/Integrals in Fractional Calculus

1.4.1.1 Riemann–Liouville

$$_aD_t^\alpha f(t) = \frac{1}{\Gamma(n-\alpha)} \left(\frac{d}{dt}\right)^n \int_a^t \frac{f(\tau)}{(t-\tau)^{\alpha-n+1}} d\tau$$

$$(n-1) \leq \alpha < n$$

where n is integer and α is real number.

1.4.1.2 Grunwald–Letnikov: (Differintegrals)

$$_aD_t^\alpha f(t) = \lim_{h \to 0} \frac{1}{h^\alpha} \sum_{j=0}^{[\frac{t-a}{h}]} (-1)^j \binom{\alpha}{j} f(t - jh)$$

$$\left[\frac{t-a}{h}\right] \to INTEGER$$

1.4.1.3 M. Caputo (1967)

$$_a^C D_t^\alpha f(t) = \frac{1}{\Gamma(n-\alpha)} \int_a^t \frac{f^{(n)}(\tau)}{(t-\tau)^{\alpha+1-n}} d\tau, (n-1) \le \alpha < n,$$

where n is integer and α is real number.

1.4.1.4 Oldham and Spanier (1974)

Fractional derivatives scaling property is

$$\frac{d^q f(\beta x)}{dx^q} = \beta^q \frac{d^q f(\beta x)}{d(\beta x)^q}$$

which makes it suitable for the study of scaling. This implies the study of self-similar processes, objects and distributions too.

1.4.1.5 K.S. Miller B. Ross (1993)

$$D^\alpha f(t) = D^{\alpha_1} D^{\alpha_2} \ldots D^{\alpha_n} f(t)$$
$$\alpha = \alpha_1 + \alpha_2 + \ldots + \alpha_n$$
$$\alpha_i < 1$$

This definition of sequential composition is a very useful concept for obtaining fractional derivative of any arbitrary order. The derivative operator can be any definition RL or Caputo.

1.4.1.6 Kolwankar and Gangal (1994)

Local fractional derivative is defined by Kolwankar and Gangal to explain the behavior of "continuous but nowhere differentiable" function. The other definitions for fractional derivative, described in this chapter, are 'non-local' quantities.

For $0 < q < 1$, the local fractional derivative is

$$D^q f(y) = \lim_{x \to y} \frac{d^q \left(f(x) - f(y)\right)}{d(x-y)^q}.$$

1.5 About Fractional Integration Derivatives and Differintegration

All the efforts to realize fractional differintegration are "interpolating" the operations between the two integer order operations. In the limit when the order of the operator approaches the nearest integer, the "generalized" differintegrals tend to normal integer order operations.

1.5.1 Fractional Integration Riemann–Liouville (RL)

The repeated n-fold integration is generalized by Gamma function for the factorial expression, when the integer n is real number α.

$$D^{-n} f(t) = J^n f(t) = f_n(t) = \frac{1}{(n-1)!} \int_0^t (t-\tau)^{n-1} f(\tau) d\tau$$

$$D^{-\alpha} f(t) = J^\alpha f(t) = f_\alpha(t) = \frac{1}{\Gamma(\alpha)} \int_0^t (t-\tau)^{\alpha-1} f(\tau) d\tau$$

Defining power function as

$$\phi_\alpha(t) = \frac{t^{\alpha-1}}{\Gamma(\alpha)}$$

and using the definition of convolution integral, the expression for the fractional integration can be therefore written as the convolution of the function and the power function.

$$D^{-\alpha} f(t) = \phi_\alpha(t) * f(t) = \int_0^t \phi_\alpha(t) f(t-\tau) d\tau.$$

This process is depicted in Fig. 1.2, where L is the Laplace operator and L^{-1} is the inverse Laplace operator.

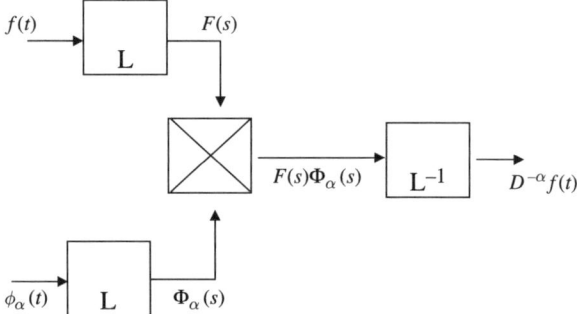

Fig. 1.2 Block diagram representation of fractional integration process by convolution

1.5.2 Fractional Derivatives Riemann–Liouville (RL) Left Hand Definition (LHD)

The formulation of this definition is:
   Select an integer m greater than fractional number α,

(i) integrate the function $(m - \alpha)$ folds by RL integration method;
(ii) differentiate the above result by m.

The expression is given as

$$D^\alpha f(t) = \frac{d^m}{dt^m}\left[\frac{1}{\Gamma(m-\alpha)}\int_0^t \frac{f(\tau)d\tau}{(t-\tau)^{\alpha+1-m}}d\tau\right]$$

Figure 1.3 gives the process block diagram, and Fig. 1.4 gives the process of differentiation of 2.3 times for a function.

1.5.3 Fractional Derivatives Caputo Right Hand Definition (RHD)

The formulation is exactly opposite to LHD.
   Select an integer m greater than fractional number α,

(i) differentiate the function m times;
(ii) integrate the above result $(m - \alpha)$-fold by RL integration method.

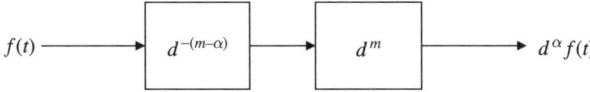

Fig. 1.3 Fractional differentiation Left Hand Definition (LHD) block diagram

1.5 About Fractional Integration Derivatives and Differintegration

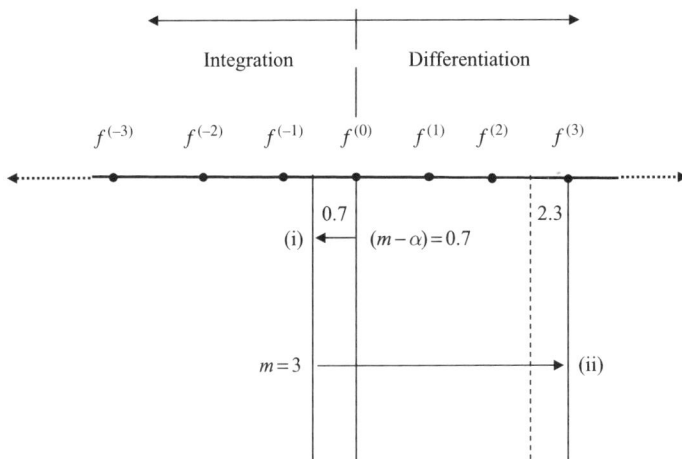

Fig. 1.4 Fractional differentiation of 2.3 times in LHD

In LHD and RHD the integer selection is made such that $(m - 1) < \alpha < m$. For example, differentiation of the function by order π will select $m = 4$. The formulation of RHD Caputo is as follows:

$$D^\alpha f(t) = \frac{1}{\Gamma(m - \alpha)} \int_0^t \frac{\frac{d^m f(t)}{dt^m}}{(t - \tau)^{\alpha+1-m}} d\tau = \frac{1}{\Gamma(m - \alpha)} \int_0^t \frac{f^{(m)}(t)}{(t - \tau)^{\alpha+1-m}} d\tau$$

Figure 1.5 gives the block diagram representation of the RHD process, and Fig. 1.6 represents graphically the RHD used for fractionally differentiating function of 2.3 times.

The definitions of Riemann-Liouville of fractional differentiation played an important role in the development of fractional calculus. However, the demands of modern science and engineering require a certain revision of the well-established pure mathematical approaches. Applied problems require definitions of fractional derivatives, allowing the utilization of physically interpretable "initial conditions" which contain $f(a)$, $f^{(1)}(a)$, $f^{(2)}(a)$ and not fractional quantities (presently unthinkable!). The RL definitions require

$$\lim_{t \to a} {}_aD_t^{\alpha-1} f(t) = b_1$$
$$\lim_{t \to a} {}_aD_t^{\alpha-2} f(t) = b_2$$

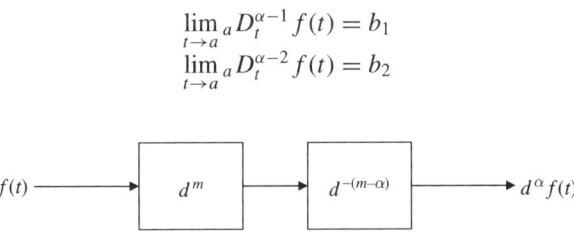

Fig. 1.5 Block diagram representation of RHD Caputo

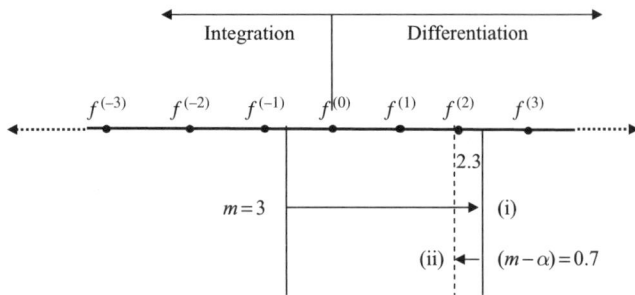

Fig. 1.6 Differentiation of 2.3 times by RHD

In spite of the fact that initial value problems with such initial conditions can be successfully solved mathematically, their solutions are practically useless because there is no known physical interpretation for such initial conditions.

RHD is more restrictive than LHD. For RL (or LHD), $f(t)$ needs to be causal, that is as long as $f(0) = 0$ for $t \leq 0$, the LHD method is workable. For RHD because $f(t)$ is first made to mth derivative, i.e., $f^{(m)}(t)$, the condition $f(0) = 0$ and $f^1 = f^2 = \ldots .f^m = 0$ is required. In mathematical world, this is vulnerable for RHD may be deliberating. For LHD

$$D^\alpha C \neq 0 = \frac{Ct^{-\alpha}}{\Gamma(1-\alpha)},$$

the derivative of constant C is not zero. This fact led to using the RL or LHD approach with lower limit of differentiation $a \to -\infty$; in physical world this poses problem. The physical meaning of this lower limit extending toward minus infinity is starting of physical process at time immemorial! In such cases, the transient effects cannot be then studied. However, making $a \to -\infty$ is a necessary abstraction for consideration of steady-state process, for example, for the study of sinusoidal analysis for steady-state fractional order system.

While today we are familiar with interpretation of the physical world with integer order differential equations, we do not (currently) have practical understanding of the world with fractional order differential equations. Our mathematical tools go beyond practical limitation of our understanding.

Therefore, still process is on to "generalize" the concepts for use in practical world.

1.5.4 Fractional Differintegrals Grunwald Letnikov (GL)

Differintegration process as described below is differentiation for positive index and integration for negative index for the differintegral (generalized) operator.

1.5 About Fractional Integration Derivatives and Differintegration

$$f^1(x) = \lim_{h \to 0} \frac{f(x+h) - f(x)}{h}$$

$$f^2(x) = \lim_{h \to 0} \frac{f^1(x+h) - f^1(x)}{h}$$

$$= \lim_{h_1 \to 0} \frac{\lim_{h_2 \to 0} \frac{f(x+h_1+h_2)-f(x+h_1)}{h_2} - \lim_{h_2 \to 0} \frac{f(x+h_2)-f(x)}{h_2}}{h_1}$$

$$h_1 = h_2 = h$$

$$f^2(x) = \lim_{h \to 0} \frac{f(x+2h) - 2f(x+h) + f(x)}{h^2}, \text{ continuing for } n \text{ times we have}$$

$$f^n(x) = D^n f(x) = \lim_{h \to 0} \frac{1}{h^n} \sum_{m=0}^{n} (-1)^m \binom{n}{m} f(x - mh).$$

$$\binom{n}{m} = \frac{n!}{m!(n-m)!}$$

This can be replaced by Gamma functions as $\frac{\Gamma(\alpha+1)}{m!\Gamma(\alpha-m+1)}$ for non-integer n, i.e., α. Therefore, differentiation in fractional order is

$$_aD^\alpha f(x) = \lim_{h \to 0} \frac{1}{h^\alpha} \sum_{m=0}^{\left[\frac{x-a}{h}\right]} (-1)^m \frac{\Gamma(\alpha+1)}{m!\Gamma(\alpha-m+1)} f(x - mh).$$

For negative α, the process will be integration.

$$\binom{-n}{m} = \frac{-n(-n-1)(-n-2)\ldots(-n-m+1)}{m!}$$

$$= (-1)^m \frac{n(n+1)(n+2)(n+3)\ldots(n+m-1)}{m!}$$

$$= (-1)^m \frac{(n+m-1)!}{m!(n-1)!} \to (-1)^m \frac{\Gamma(\alpha+m)}{m!\Gamma(\alpha)}$$

Therefore, for integration we write

$$_aD^{-\alpha} f(x) = \lim_{h \to 0} h^\alpha \sum_{m=0}^{\left[\frac{x-a}{h}\right]} \frac{\Gamma(\alpha+m)}{m!\Gamma(\alpha)} f(x - mh).$$

The part $[(x - a)/h]$ is integer part (floor function). That is, the upper limit of the summation is the integer part of the fraction. Are RL, GL, RHD (Caputo) and LHD, equivalent? The answer is, "yes."

1.5.5 Composition and Property

In this book, the symbols for fractional differintegration have been standardized as follows.

$_cD_t^q f(t)$ represents initialized qth order differintegration of $f(t)$ from start point c to t.
$_cd_t^q f(t)$ represents un-initialized generalized (or fractional) qth order differintegral.

This is also same as

$$\frac{d^q f(t)}{[d(t-c)]^q} \equiv {_cd_t^q} f(t),$$

shifting the origin of function at start of the point from where differintegration starts. This un-initialized operator can also be short, formed as $d^q f(t)$.

The index $q > 0$ is differentiation, and the index $q < 0$ is integration process. For q as integer, the process is integer order classical differentiation and integration.

Miller and Ross (1993), with sequential fractional derivatives, tried to give formal properties and to have composition methods for generalized differintegrals. Decomposition $_aD_t^\alpha y(t) = {_aD_t^m} {_aD_t^{\alpha-m}} y(t)$ and also to some extent the index commutation (under certain conditions) $D^{-\alpha} D^{-\beta} = D^{-(\alpha+\beta)} = D^{-\beta} D^{-\alpha}$ are well true for fractional integration. But fractional derivatives do not commute always, i.e., $D^\alpha D^\beta \neq D^{\beta+\alpha} \neq D^{\alpha+\beta}$ (except at zero initial conditions). Integer operator (m) commutes with fractional operator (α), i.e., $D^m D^\alpha = D^{m+\alpha}$, some of the basic composition properties. The desirable properties of fractional derivatives and integrals are the following:

a. If $f(z)$ is an analytical function of z, then its fractional derivatives $_0D_z^\alpha f(z)$ is an analytical function of z and α.
b. For $\alpha = n$, where n is integer, the operation $_0D_z^\alpha f(z)$ gives the same result as the classical differentiation or integration of integer order n.
c. For $\alpha = 0$, the operation $_0D_z^\alpha f(z)$ is identity operator, i.e., $_0D_z^0 f(z) = f(z)$
d. Fractional differentiation and fractional integration are linear operations:
$$_0D_z^\alpha a f(z) + {_0D_z^\alpha} b g(z) = a_0 D_z^\alpha f(z) + b_0 D_z^\alpha g(z)$$
e. The additive index law, $_0D_z^\alpha {_0D_z^\beta} f(z) = {_0D_z^\beta} {_0D_z^\alpha} f(z) = {_0D_z^{\alpha+\beta}} f(z)$, holds under some reasonable constraints on the function $f(z)$.

The above desirable properties are valid under causality; that is, the function is differintegrated at the start point of the function itself (with initialization function being zero).

1.5.6 Fractional Derivative for Some Standard Function

Table 1.1 lists Riemann–Liouvelle fractional derivatives of some functions, which are used very often. In most cases, the order of differentiation α may be a real number, so replacing it with $-\alpha$ gives Riemann–Liouville fractional integral. Table 1.1 can be used to find Grunwald–Letnikov, fractional derivatives, Caputo fractional derivatives, and Miller–Ross sequential fractional derivatives. In such cases, α should be taken between 0 and 1, and Riemann–Liouvelli fractional derivatives should be properly combined (composed) with integer order derivatives, with considered definition (of composition). Table 1.1 gives the RL derivatives with lower terminal at 0, and Table 1.2 gives the fractional RL derivatives with lower terminal

Table 1.1 RL derivative with lower terminal 0 i.e., $_0D_t^\alpha f(t)$ for $t > 0$

function f (t)	$_0D_t^\alpha f(t)$. fractional derivative
$H(t)$	$\dfrac{t^{-\alpha}}{\Gamma(1-\alpha)}$
$H(t-a)$	$\begin{cases} \dfrac{(t-a)^{-\alpha}}{\Gamma(1-\alpha)}, (t > a) \\ 0, (0 \le t \le a) \end{cases}$
$H(t-a)f(t)$	$\begin{cases} {_aD_t^\alpha f(t)}, (t > a) \\ 0, (0 \le t \le a) \end{cases}$
$\delta(t)$	$\dfrac{t^{-\alpha-1}}{\Gamma(-\alpha)}$
$\delta^{(n)}(t)$	$\dfrac{t^{-\alpha-n-1}}{\Gamma(-\alpha-n)}$
$\delta^{(n)}(t-a)$	$\begin{cases} \dfrac{(t-a)^{-\alpha-n-1}}{\Gamma(-n-\alpha)}, (t > a) \\ 0, (0 \le t \le a) \end{cases}$
t^v	$\dfrac{\Gamma(v+1)}{\Gamma(v+1-\alpha)} t^{v+\alpha} \quad v > -1$
$e^{\lambda t}$	$t^{-\alpha} E_{1,1-\alpha}(\lambda t)$
$\cosh(\sqrt{\lambda} t)$	$t^{-\alpha} E_{2,1-\alpha}(\lambda t^2)$
$\dfrac{\sinh(\sqrt{\lambda} t)}{\sqrt{\lambda} t}$	$t^{1-\alpha} E_{2,2-\alpha}(\lambda t^2)$
$\ln(t)$	$\dfrac{t^{-\alpha}}{\Gamma(1-\alpha)} (\ln(t) + \psi(1) - \psi(1-\alpha))$
$t^{\beta-1} \ln(t)$	$\dfrac{\Gamma(\beta) t^{\beta-\alpha-1}}{\Gamma(\beta-\alpha)} (\ln(t) + \psi(\beta) - \psi(\beta-\alpha))$
$t^{\beta-1} E_{\mu,\beta}(\lambda t^\mu)$	$t^{\beta-\alpha-1} E_{\mu,\beta-\alpha}(\lambda t^\mu)$

Table 1.2 RL derivative with lower terminal at $-\infty$ i.e., $_{-\infty}D_t^\alpha f(t)$

function $f(t)$	$_{-\infty}D_t^\alpha f(t)$ derivative
$H(t-a)$	$\begin{cases} \dfrac{(t-a)^{-\alpha}}{\Gamma(1-\alpha)}, (t>a) \\ 0, (t \le a) \end{cases}$
$H(t-a)f(t)$	$\begin{cases} {}_aD_t^\alpha f(t), (t>a) \\ 0, (t \le a) \end{cases}$
$e^{\lambda t}$	$\lambda^\alpha e^{\lambda t}$
$e^{\lambda t + \mu}$	$\lambda^\alpha e^{\lambda t + \mu}$
$\sin \lambda t$	$\lambda^\alpha \sin\left(\lambda t + \dfrac{\pi\alpha}{2}\right)$
$\cos \lambda t$	$\lambda^\alpha \cos\left(\lambda t + \dfrac{\pi\alpha}{2}\right)$
$e^{\lambda t} \sin \mu t$	$r^\alpha e^{\lambda t} \sin(\mu t + \alpha\varphi)\ r = \sqrt{\lambda^2 + \mu^2}\quad \tan\varphi = \dfrac{\mu}{\lambda}\quad (\lambda, \mu > 0)$
$e^{\lambda t} \cos \mu t$	$r^\alpha e^{\lambda t} \cos(\mu t + \alpha\varphi)\ r = \sqrt{\lambda^2 + \mu^2}\quad \tan\varphi = \dfrac{\mu}{\lambda}\quad (\lambda, \mu > 0)$

at $-\infty$. In the list, $H(t)$ is unit step Heaviside function. E is Mittag-Leffler function. These tables give a feel of how fractional differintegration will look like in analytical expressions.

1.6 Solution of Fractional Differential Equations

Fractional differential equations appear in several physical systems. Solution to these is no more rigorous than its integer order counterpart. The Laplace transformation technique is very popular, though several analytical approaches do exist. Numerical evaluation with "short-memory principle" is one among them popular for computer programing and numerical regression. Mellin transform, power series expansion method approach using fractional Green's function, Babenko's symbolic method, orthogonal polynomial method, Reisz fractional potential method, method with Wright's function, and finite-part integral method are some of the mathematician's tool for obtaining the fractional differintegrals and solution of fractional differential equation. However, in this book, only the Laplace transformation is considered as it is easily understood and being popular among engineers and scientists.

1.7 A Thought Experiment

From an aircraft, we can see the city roads and observe the vehicular traffic movement. The vehicle seems to move in a straight line. Therefore, as an observer, we draw the velocity curve by simple one-order integer derivative of displacement and

Fig. 1.7 Macroscopic and microscopic view of moving vehicles on road

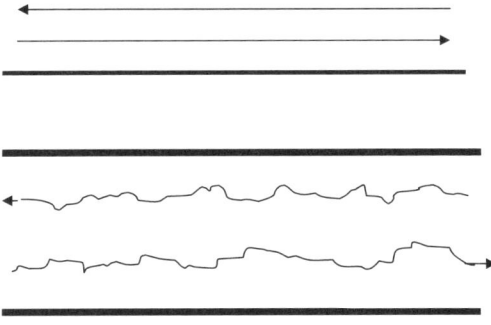

find that it maps a straight line. In Fig. 1.7, the pair of straight line gives the velocity trajectory of the upstream vehicle and downstream vehicle, as observed in macroscopic scale.

The same vehicle when looked with enlarged view tells us its continuous movement but to avoid road heterogeneity it travels in zigzag fashion. The curve in the lower frame of Fig. 1.7 maps this picture. Here the scale is enlarged. The velocities for upstream and downstream vehicles are not pair of straight lines, but follow a continuous, nowhere differentiable curve. So will the dx/dt give the true picture of velocity or will it be $d^{1+\alpha}x/dt^{1+\alpha}$, where $0 < \alpha < 1$, give the representation of the actual zigzag pattern is the thought experiment. Now the question about the dimensions of velocity, in the thought experiment when defined as fractional derivative of displacement, is the matter of another thought. In the present understanding, as per uniform time scales, the quantity dx/dt is velocity, and d^2x/dt^2 is the acceleration; however, the quantification of $d^{1.23}x/dt^{1.23}$ is hard to visualize. This fractional differentiation is in between velocity and acceleration, perhaps a velocity in some transformed time scale, which is non-uniform-enriching thought for physical understanding of fractional quantities. The nature of zigzag pattern shown is somewhat called fractal curve, actually a continuous and nowhere differentiable function. The relation of fractal dimensions and fractional calculus is an evolving field of science at present. The macroscopic view presented above gives a thought of explanation of discontinuity and singularity formations in nature, in classical integer order calculus. Can fractional calculus be an aid for explanation of discontinuity formation and singularity formation is an enriching thought experiment.

1.8 Quotable Quotes About Fractional Calculus

Expressed differently we may say that nature works with fractional derivatives.

We may express our concepts in Newtonian terms if we find it convenient, but if we do so, we must realize that we have made translation into a language, which is foreign to the system we are studying.

All systems need a fractional time derivative in the equation describing them. System having memory of all earlier events is thus necessary to include this record of earlier events to predict the future. Conclusion is obvious and unavoidable: "Dead matter has memory."

Fractional calculus is the calculus of twenty-first century.

1.9 Concluding Comments

This field of science is evolving, and particularly as per author's intuition, this calculus will be the language of twenty-first century for physical system description and controls. In this chapter, observation points toward evolving nature of the science of fractional calculus definitions, though born 300 hundred years ago. The "ifs and buts," related to fractional calculus as today, is due to our own limitation of understanding. This will have a clearer picture tomorrow when products based on this subject will be used in the industry. This introduction chapter has given the thought that there is a wonderful universe of mathematics staying within the boundary of one complete differentiation and one complete integration. The science maturity will absorb the richness in this fractional calculus may be in coming years of twenty-first century. The chapter gives the idea that this fractional calculus is as rigorous as its counterpart classical integer order differentiation and integration, with subject's richness for scientific research for future.

Chapter 2
Functions Used in Fractional Calculus

2.1 Introduction

This chapter presents a number of functions that have been found to be useful in providing solutions to the problems of fractional calculus. The base function is the Gamma function, which generalizes the factorial expression, used in multiple differentiation and repeated integrations, in integer order calculus. The Mittag-Leffler function is the basis function of fractional calculus, as the exponential function is to the integer order calculus. Several modifications of the Mittag-Leffler functions are introduced which are developed since 1903, for study of the fractional calculus.

2.2 Functions for the Fractional Calculus

2.2.1 Gamma Function

One of the basic functions of the fractional calculus is Euler's Gamma function. This function generalizes the factorial $n!$ and allows n to take non-integer values.

2.2.1.1 Definition of the Gamma Function

$\Gamma(z) = \int_0^\infty e^{-t} t^{z-1} dt$, which converges in the right half of the complex plane $\Re e(z) > 0$.

Considering z to be real number, the above statement implies that Gamma function is defined continuously for positive real values of z.

2.2.1.2 Basic Properties of Gamma Function

$$\Gamma(z+1) = z\Gamma(z)$$

$$\Gamma(z+1) = \int_0^\infty e^{-t} t^{(z+1)-1} dt = \int_0^\infty e^{-t} t^z dt$$

$$= \left[-e^{-t} t^z\right]_{t=0}^{t=\infty} + z \int_0^\infty e^{-t} t^{z-1} dt$$

$$= z\Gamma(z)$$

The above equation is obtained by integration by parts. Obviously $\Gamma(1) = 1$, and using the above property, we obtain values for $z = 1, 2, 3, \ldots$

$$\Gamma(2) = 1.\Gamma(1) = 1!$$
$$\Gamma(3) = 2.\Gamma(2) = 2!$$
$$\Gamma(4) = 3.\Gamma(3) = 3!$$
$$\ldots\ldots\ldots\ldots\ldots\ldots$$
$$\Gamma(n+1) = n.\Gamma(n) = n.(n-1)! = n!$$

The above property is valid for positive values of z. Another important property of the Gamma function is that it has simple poles at $z = 0, -1, -2, -3, \ldots$

The proof is explained by splitting the function into two intervals, as indicated below:

$$\Gamma(z) = \int_0^1 e^{-t} t^{z-1} dt + \int_1^\infty e^{-t} t^{z-1} dt$$

The first integral can be evaluated by using series expansion for the exponential function. If $\Re e(z) = x > 0$, then $\Re e(z+k) = x + n > 0$ and thus $t^{z+k}\big|_{t=0} = 0$. Therefore,

$$\int_0^1 e^{-t} t^{z-1} dt = \int_0^1 \sum_{k=0}^\infty \frac{(-t)^k}{k!} t^{z-1} dt = \sum_{k=0}^\infty \frac{(-1)^k}{k!} \int_0^1 t^{k+z-1} dt = \sum_{k=0}^\infty \frac{(-1)^k}{k!(k+z)}.$$

The second integral may be represented as an "entire- function"

2.2 Functions for the Fractional Calculus

$$\varphi(z) = \int_1^\infty e^{-t} t^{z-1} dt$$

$$\Gamma(z) = \sum_{k=0}^\infty \frac{(-1)^k}{k!} \frac{1}{k+z} + \varphi(z)$$

$$= \varphi(z) + \frac{(-1)^0}{0!} \frac{1}{0+z} + \frac{(-1)^1}{1!} \frac{1}{1+z} + \frac{(-1)^2}{2!} \frac{1}{2+z} + \ldots.$$

thus clearly indicating simple poles at $0, -1, -2, -3\ldots$, this means that at negative integer points, the Gamma function asymptotically approaches infinity and is discontinuous at those negative integer values.

2.2.1.3 Gamma Function Defined by Limit

The gamma function can also be represented as a limit as follows:

$$\Gamma(z) = \lim_{n\to\infty} \frac{n! \cdot n^z}{z(z+1)\ldots(z+n)}.$$

Here, we initially assume the right half plane $\Re e(z) > 0$, or in case of real number positive values.

Let us introduce an auxiliary function to prove this part:

$$f_n = \int_0^n \left(1 - \frac{t}{n}\right)^n t^{z-1} dt.$$

Substitute $\tau = t/n$ and then performing integration by parts we get the following:

$$f_n(z) = n^z \int_0^1 (1-\tau)^n \tau^{z-1} d\tau$$

$$= \frac{n^z}{z} n \int_0^1 (1-\tau)^{n-1} \tau^z d\tau$$

$$= \frac{n^z n!}{z(z+1)\ldots(z+n-1)} \int_0^1 \tau^{z+n-1} d\tau$$

$$= \frac{n^z n!}{z(z+1)\ldots(z+n-1)(z+n)}$$

Taking into account the well-known $\lim_{n\to\infty} \left(1 - \frac{t}{n}\right)^n = e^{-t}$, we expect the following:

$$\lim_{n\to\infty} f_n(z) = \lim_{n\to\infty} \int_0^n \left(1 - \frac{t}{n}\right)^n t^{z-1} dt = \int_0^\infty e^{-t} t^{z-1} dt = \Gamma(z)$$

Incomplete Gamma function is defined in two ways. In both the definitions, the integrand is same, but limits of integration is different. The Upper-incomplete Gamma function is defined as

$$\Gamma(z, x) = \int_x^\infty t^{z-1} e^{-t} dt,$$

and the Lower-incomplete Gamma function is defined as

$$\gamma(z, x) = \int_0^x t^{z-1} e^{-t} dt,$$

and in both the cases, x is real and $x \geq 0$, and z is complex with $\Re e(z) > 0$. Some of the properties of incomplete Gamma functions are

$$\Gamma(z) = \Gamma(z, x) + \gamma(z, x)$$
$$\Gamma(z+1, x) = z\Gamma(z, x) + x^z e^{-x}$$
$$\gamma(z+1, x) = z\gamma(z, x) - x^z e^{-x}$$

For integer $n = z$,

$$\Gamma(n, x) = (n-1)! e^{-x} \sum_{k=0}^{n-1} \frac{x^k}{k!}$$
$$\Gamma(n, 0) = \Gamma(n) = (n-1)!$$
$$\gamma(n, x) \to \Gamma(n), \lim x \to \infty$$
$$\Gamma(1, x) = e^{-x}$$
$$\gamma(1, x) = 1 - e^{-x}$$

The incomplete Gamma function is used in obtaining fractional differentiation and fractional integration of periodic functions, used as sinusoidal response studies of fractional operators.

2.2.2 Mittag-Leffler Function

In the integer order calculus equations, the exponential function exp(z) plays an important role. Similarly in the fractional order calculus, the Mittag-Leffler function plays the important part.

2.2 Functions for the Fractional Calculus

For this new function $E_q[az]$, $q > 0$, Mittag-Leffler considered the parameter a to be a complex number, such as $a = |a|\exp(j\phi)$. As he studied this function, it became apparent that this function is either stable (decays to zero) or unstable (goes to infinity) as z increases, depending upon how the parameter a and q are chosen. The result was that the function remained bounded for increasing z if $|\phi| \geq q\frac{\pi}{2}$.

2.2.2.1 One-Parameter Mittag-Leffler Function

It is defined as

$$E_\alpha(z) = \sum_{k=0}^{\infty} \frac{z^k}{\Gamma(\alpha k + 1)}$$

The expanded form is the infinite series that is as follows:

$$E_\alpha(z) = 1 + \frac{z}{\Gamma(\alpha + 1)} + \frac{z^2}{\Gamma(2\alpha + 1)} + \frac{z^3}{\Gamma(3\alpha + 1)} + \cdots$$

This function was introduced by Mittag-Leffler in 1903.

2.2.2.2 Two-Parameter Mittag-Leffler Functions

Two-parameter Mittag-Leffler function plays a very important role in fractional calculus. This function type was introduced by R. P. Agarwal and Erdelyi in 1953–1954.

The two-parameter function is defined as follows:

$$E_{\alpha,\beta}(z) = \sum_{k=0}^{\infty} \frac{z^k}{\Gamma(\alpha k + \beta)} \quad (\alpha > 0, \beta > 0)$$

$E_{\alpha,1}(z) = \sum_{k=0}^{\infty} \frac{z^k}{\Gamma(\alpha k + 1)} \equiv E_\alpha(z)$ is one-parameter Mittag-Leffler function.

The following identities follow from the definition:

$$E_{1,1}(z) = \sum_{k=0}^{\infty} \frac{z^k}{\Gamma(k+1)} = \sum_{k=0}^{\infty} \frac{z^k}{k!} = e^z$$

$$E_{1,2}(z) = \sum_{k=0}^{\infty} \frac{z^k}{\Gamma(k+2)} = \sum_{k=0}^{\infty} \frac{z^k}{(k+1)!} = \frac{1}{z}\sum_{k=0}^{\infty} \frac{z^{k+1}}{(k+1)!} = \frac{e^z - 1}{z}$$

$$E_{1,3}(z) = \sum_{k=0}^{\infty} \frac{z^k}{\Gamma(k+3)} = \sum_{k=0}^{\infty} \frac{z^k}{(k+2)!} = \frac{1}{z^2}\sum_{k=0}^{\infty} \frac{z^{k+2}}{(k+2)!} = \frac{e^z - 1 - z}{z^2}$$

The above equation have the general form as follows:

$$E_{1,m}(z) = \frac{1}{z^{m-1}}\left(e^z - \sum_{k=0}^{m-2}\frac{z^k}{k!}\right)$$

The trigonometric and hyperbolic functions are also manifestations of the two-parameter Mittag-Leffler function, which is indicated below:

$$E_{2,1}(z^2) = \sum_{k=0}^{\infty}\frac{z^{2k}}{\Gamma(2k+1)} = \sum_{k=0}^{\infty}\frac{z^{2k}}{(2k)!} = \cosh(z),$$

$$E_{2,2}(z^2) = \sum_{k=0}^{\infty}\frac{z^{2k}}{\Gamma(2k+2)} = \frac{1}{z}\sum_{k=0}^{\infty}\frac{z^{2k+1}}{(2k+1)!} = \frac{\sinh(z)}{z}$$

Generalized hyperbolic function of order n is represented below:

$$h_r(z,n) = \sum_{k=0}^{\infty}\frac{z^{nk+r-1}}{(nk+r-1)!} = z^{r-1}E_{n,r}(z^n), \ldots (r = 1,2,3,\ldots,n)$$

and the generalized trigonometric function of order n is also represented below:

$$k_r(z,n) = \sum_{m=0}^{\infty}\frac{(-1)^m z^{nm+r-1}}{(nm+r-1)!} = z^{r-1}E_{n,r}(-z^n), \ldots (r = 1,2,3,\ldots,n)$$

Mathematical handbooks describe $erfc(z)$ as follows:
The error function is defined as

$$erf(z) = \frac{2}{\sqrt{\pi}}\int_0^z e^{-t^2}dt$$

and is represented by series as

$$erf(z) = \frac{2}{\sqrt{\pi}}\sum_{n=0}^{\infty}\frac{(-1)^n z^{2n+1}}{(2n+1)n!} = \frac{2}{\sqrt{\pi}}\left(z - \frac{z^3}{3} + \frac{z^5}{10} - \frac{z^7}{42} + \frac{z^9}{216} + \ldots\right)$$

The complimentary error function is defined as

$$erfc(z) = 1 - erf(z) = 1 - \frac{2}{\sqrt{\pi}}\int_0^z e^{-t^2}dt = \frac{2}{\sqrt{\pi}}\int_z^{\infty} e^{-t^2}dt.$$

2.2 Functions for the Fractional Calculus

The series asymptotic expansion of complimentary error function is

$$erfc(z) = \frac{e^{-z^2}}{z\sqrt{\pi}}\left[1 + \sum_{n=1}^{\infty}(-1)^n \frac{1.3.5\ldots(2n-1)}{(2z^2)^n}\right]$$

$$= \frac{e^{-z^2}}{z\sqrt{\pi}}\left[1 + \sum_{n=1}^{\infty}(-1)^n \frac{(2n)!}{n!(2z)^{2n}}\right]$$

2.2.2.3 Variants of Mittag-Leffler Function

$$\xi_t(\nu, a) = t^\nu \sum_{k=0}^{\infty} \frac{(at)^k}{\Gamma(\nu + k + 1)} = t^\nu E_{1,\nu+1}(at)$$

This function is important for solving fractional differential equations.

$$\Im_\alpha(\beta, t) = t^\alpha \sum_{k=0}^{\infty} \frac{\beta^k t^{k(\alpha+1)}}{\Gamma(\{k+1\}\{\alpha+1\})} = t^\alpha E_{\alpha+1,\alpha+1}(\beta t^{\alpha+1})$$

This function is called Rabotnov function, and a special variant too.

$$Sc_\alpha(z) = \sum_{n=0}^{\infty} \frac{(-1)^n z^{(2-\alpha)n+1}}{\Gamma(\{2-\alpha\}n + 2)} = z E_{2-\alpha,2}(-z^{2-\alpha})$$

is the fractional sine function form-I.

$$Cs_\alpha(z) = \sum_{n=0}^{\infty} \frac{(-1)^n z^{(2-\alpha)n}}{\Gamma(\{2-\alpha\}n + 1)} = E_{2-\alpha,1}(-z^{2-\alpha})$$

is the fractional cosine function form-I.

$$\sin_{\lambda,\mu}(z) = \sum_{k=0}^{\infty} \frac{(-1)^k z^{2k+1}}{\Gamma(2\mu k + 2\mu - \lambda + 1)} = z E_{2\mu,2\mu-\lambda+1}(-z^2)$$

is the fractional sine function form-II, and

$$\cos_{\lambda,\mu}(z) = \sum_{k=0}^{\infty} \frac{(-1)^k z^{2k}}{\Gamma(2\mu k + \mu - \lambda + 1)} = E_{2\mu,\mu-\lambda+1}(-z^2)$$

is the fractional cosine function form-II.

Generalization of the Mittag-Leffler function to two variables was suggested and were further extended by Srivastava to the following type of symmetric form.

$$\xi_{\alpha,\beta,\lambda,\mu}^{\nu,\sigma} = \sum_{m=0}^{\infty}\sum_{n=0}^{\infty} \frac{x^{m+\frac{\beta(\nu n+1)-1}{\alpha}} y^{n+\frac{\mu(\sigma m+1)-1}{\lambda}}}{\Gamma(m\alpha+(\nu n+1)\beta)\Gamma(n\lambda+(\sigma m+1)\mu)}$$

Several manifestations including several variables representing Mittag-Leffler have been made for multi-dimensional studies on fractional calculus.

2.2.2.4 Laplace Transforms of Mittag-Leffler Function

The following expressions give some identities for Laplace transforms pairs of Mittag-Leffler functions

$$t^{\alpha k+\beta-1} E_{\alpha,\beta}^{(k)}(at^\alpha) \leftrightarrow \frac{s^{\alpha-\beta} k!}{(s^\alpha - a)^{k+1}}$$

here

$$E_{\alpha,\beta}^{(k)} = \frac{d^{(k)}}{dt^{(k)}} E_{\alpha,\beta}$$

For $k > 0$ the operation is differentiation of Mittag-Leffler function, and for $k < 0$ the operation is integration of Mittag- Leffler function.

$$\beta = 1, k = 0, \rightarrow E_{\alpha,1}(at^\alpha) \rightarrow E_\alpha(at^\alpha) \leftrightarrow \frac{s^{\alpha-1}}{s^\alpha - a}$$

$$E_\alpha(-\lambda t^\alpha) \leftrightarrow \frac{s^{\alpha-1}}{s^\alpha + \lambda}$$

$$E_\alpha(-t^\alpha) \leftrightarrow \frac{s^{\alpha-1}}{s^\alpha + 1}$$

$$E_\alpha(\lambda t^\alpha) \leftrightarrow \frac{s^\alpha}{s^\alpha - \lambda}$$

$$\frac{d}{dt} E_\alpha(-\lambda t^\alpha) \leftrightarrow \frac{\lambda}{s^\alpha + \lambda} = -\left(s\frac{s^{\alpha-1}}{s^\alpha + \lambda} - 1\right)$$

$$\frac{d}{dt} E_\alpha(-t^\alpha) \leftrightarrow \frac{1}{s^\alpha + 1} = -\left(s\frac{s^{\alpha-1}}{s^\alpha + 1} - 1\right)$$

$$\frac{d^{-1}}{dt^{-1}} E_\alpha(-t^\alpha) \leftrightarrow \frac{s^{\alpha-2}}{s^\alpha + 1} = \frac{1}{s}\frac{s^{\alpha-1}}{s^\alpha + 1}$$

$$\frac{d^{-k}}{dt^{-k}} E_\alpha(-t^\alpha) \leftrightarrow \frac{s^{\alpha-k-1}}{s^\alpha + 1} = \frac{1}{s^k}\frac{s^{\alpha-1}}{s^\alpha + 1}$$

2.2 Functions for the Fractional Calculus

2.2.3 Agarwal Function

The Mittag-Leffler function is generalized by Agarwal in 1953. This function is particularly interesting to the fractional order system theory due to its Laplace transform given by Agarwal. The function is defined as follows:

$$E_{\alpha,\beta}(t) = \sum_{m=0}^{\infty} \frac{t^{\left(m+\frac{\beta-1}{\alpha}\right)}}{\Gamma(\alpha.m+\beta)}$$

$$L\{E_{\alpha,\beta}(t^\alpha)\} = \frac{s^{\alpha-\beta}}{s^\alpha - 1}$$

2.2.4 Erdelyi's Function

Erdelyi (1954) has studied the generalization of Mittag-Leffler function as

$$E_{\alpha,\beta}(t) = \sum_{m=0}^{\infty} \frac{t^m}{\Gamma(\alpha.m+\beta)}, \alpha, \beta > 0$$

where the powers of t are integers.

2.2.5 Robotnov–Hartley Function

To effect the direct solution of the fundamental linear fractional order differential equations, the following function was introduced by Robotnov and Hartley (1998)

$$F_q(-a, t) = t^{q-1} \sum_{n=0}^{\infty} \frac{(-a)^n t^{nq}}{\Gamma(nq+q)}, q > 0$$

This function is the "impulse response" of the fundamental fractional differential equation and is used by control system analysis to obtain the forced or the initialized system reaction.

2.2.6 Miller–Ross Function

In 1993, Miller and Ross introduced a function as the basis of the solution of fractional order initial value problem. It is defined as the vth integral of the exponential function, that is,

$$E_t(v, a) = \frac{d^{-v}}{dt^{-v}} e^{at} = t^v \sum_{k=0}^{\infty} \frac{(at)^k}{\Gamma(v+k+1)}$$

2.2.7 Generalized R Function and G Function

It is of significant usefulness to develop a generalized function which when fractionally differentiated or integrated (differintegrated) by any order returns itself. Like exponential, trigonometric, hyperbolic functions of integer order calculus, the definitions of such generalized Mittag-Leffler functions are important in fractional calculus. In an earlier section, some variants of Mittag-Leffler are noted; here more generalized R function and G function are introduced.

$$R_{q,v}[a, c, t] = \sum_{n=0}^{\infty} \frac{(a)^n (t-c)^{(n+1)q - 1 - v}}{\Gamma\{(n+1)q - v\}} \equiv R_{q,v}[a, t-c]$$

Here t is independent variable and c is the lower limit of fractional differintegration. Our interest in this function will be normally for the range $t > c$.

The Laplace transforms of R function are

$$R_{q,v}(a, 0, t) \leftrightarrow \frac{s^v}{s^q - a}$$

$$R_{q,v}(a, c, t) \leftrightarrow \frac{e^{-cs} s^v}{s^q - a}$$

2.2.7.1 Relation to Elementary Functions

$$R_{1,0}(a, 0, t) = e^{at}$$

$$a R_{2,0}(-a^2, 0, t) = a \left\{ t - \frac{a^2 t^3}{3!} + \frac{a^4 t^5}{5!} - \ldots \right\} = \sin(at)$$

$$R_{2,1}(-a^2, 0, t) = \left\{ 1 - \frac{a^2 t^2}{2!} + \frac{a^4 t^4}{4!} - \ldots \right\} = \cos(at)$$

$$a R_{2,0}(a^2, 0, t) = \sinh(at)$$

$$R_{2,1}(a^2, 0, t) = \cosh(at)$$

$$R_{1,0}(a, 0, x) = e^{ax}$$

2.2 Functions for the Fractional Calculus

2.2.7.2 Relationship of *R* Function to Other Generalized Function

Mittag-Leffler function:

$$L\{E_q[-at^q]\} = \frac{1}{s}\left[\frac{s^q}{s^q+a}\right] = \frac{s^{q-1}}{s^q+a}, q > 0$$

$$E_q(-at^q) \leftrightarrow \frac{s^{q-1}}{s^q+a} \leftrightarrow R_{q,q-1}(-a,0,t)$$

Agarwal function:

$$E_{q,p}(t^q) \leftrightarrow \frac{s^{q-p}}{s^q-1} \leftrightarrow R_{q,q-p}(1,0,t)$$

Erdelyi's function:

$$t^{1-\beta}E_{q,\beta}(t^q) = R_{q,q-\beta}(1,0,t) = t^{1-\beta}\sum_{n=0}^{\infty}\frac{t^{nq}}{\Gamma(nq+1)}$$

Robotnov and Hartley function:

$$F_q(-a,t) \leftrightarrow \frac{1}{s^q+a} \leftrightarrow R_{q,0}(-a,0,t) = \sum_{n=0}^{\infty}\frac{(-a)^n t^{(n+1)q-1}}{\Gamma(\{n+1\}q)}$$

$$L\{F_q[a,t]\} = \frac{1}{s^q-a}, q > 0$$

$$L\{E_q[-at^q]\} = \frac{1}{s}[s^q - L\{F_q[-a,t]\}]$$

$$_0d_t^{q-1}F_q[a,t] = E_q[at^q]$$

$$L^{-1}\left\{\frac{1}{s(s^q+a)}\right\} = \frac{1}{a}[1 - E_q(-at^q)] = {_0d_t^{-q}}E_q[at^q]$$

$$L^{-1}\left\{\frac{s^q}{s^q+a}\right\} = {_0d_t^q}F_q[-a,t] = {_0d_t^1}E_q[-at^q] = L^{-1}\left\{1 - \frac{a}{s^q+a}\right\}$$

$$= \delta(t) - aF_q[-a,t]$$

Miller and Ross function:

$$E_t(v,a) \leftrightarrow \frac{s^{-v}}{s-a} \leftrightarrow R_{1,-v}(a,0,t) = \sum_{n=0}^{\infty}\frac{(a)^n t^{n+v}}{\Gamma(n+v+1)}$$

2.2.7.3 Further Generalized Function (G Function)

$$G_{q,v,r}(at) = \sum_{j=0}^{\infty} \frac{\{(-r)(-1-r)\ldots(1-j-r)\}(-a)^j t^{(r+j)q-v-1}}{\Gamma(1+j)\Gamma(\{r+j\}q-v)}$$

function	time expression $f(t)$	Laplace transform $F(s)$
Mittag-Leffler	$E_q(at^q) = \sum_{n=0}^{\infty} \frac{a^n t^{nq}}{\Gamma(nq+1)}$	$\dfrac{s^q}{s(s^q-a)}$
Agarwal	$E_{\alpha,\beta}(t^\alpha) = \sum_{m=0}^{\infty} \dfrac{t^{\left(m+\frac{\beta-1}{\alpha}\right)\alpha}}{\Gamma(\alpha m+\beta)}$	$\dfrac{s^{\alpha-\beta}}{s^\alpha - 1}$
Erdelyi	$E_{\alpha,\beta}(t) = \sum_{m=0}^{\infty} \dfrac{t^m}{\Gamma(\alpha m+\beta)}$	$\sum_{m=0}^{\infty} \dfrac{\Gamma(m+1)}{\Gamma(\alpha m+\beta)s^{m+1}}$
Robotnov–Hartley	$F_q(a,t) = \sum_{n=0}^{\infty} \dfrac{a^n t^{(n+1)q-1}}{\Gamma(\{n+1\}q)}$	$\dfrac{1}{s^q-a}$
Miller–Ross	$E_t(v,a) = \sum_{k=0}^{\infty} \dfrac{a^k t^{k+v}}{\Gamma(v+k+1)}$	$\dfrac{s^{-v}}{s-a}$
Generalized R	$R_{q,v}(a,t) = \sum_{n=0}^{\infty} \dfrac{a^n t^{(n+1)q-1-v}}{\Gamma(\{n+1\}q-v)}$	$\dfrac{s^v}{s^q-a}$
Generalized G	$G_{q,v,r}(at) = \sum_{j=0}^{\infty} \dfrac{\{(-r)(-1-r)\ldots(1-j-r)\}(-a)^j t^{(r+j)q-v-1}}{\Gamma(1+j)\Gamma(\{r+j\}q-v)}$	$\dfrac{s^v}{(s^q-a)^r}$

2.3 List of Laplace and Inverse Laplace Transforms Related to Fractional Calculus

Laplace transform $F(s)$	time expression $f(t)$
$\dfrac{s^{\alpha-1}}{s^\alpha \mp \lambda}$, $\Re(s) > |\lambda|^{1/\alpha}$	$E_{\alpha,1}(\pm\lambda t^\alpha)$
$\dfrac{k! s^{\alpha-\beta}}{(s^\alpha \mp \lambda)^{k+1}}$, $\Re(s) > |\lambda|^{1/\alpha}$	$t^{\alpha k+\beta-1} E_{\alpha,\beta}^{(k)}(\pm\lambda, t^\alpha)$
$\dfrac{k!}{(\sqrt{s} \mp \lambda)^{k+1}}$, $\Re(s) > \lambda^2$	$t^{\frac{k-1}{2}} E_{\frac{1}{2},\frac{1}{2}}^{(k)}(\pm\lambda\sqrt{t})$
$\dfrac{1}{s^\alpha}$	$\dfrac{t^{\alpha-1}}{\Gamma(\alpha)}$
$\arctan \dfrac{k}{s}$	$\dfrac{1}{t}\sin(kt)$
$\log \dfrac{s^2-a^2}{s^2}$	$\dfrac{2}{t}(1-\cosh at)$

2.3 List of Laplace and Inverse Laplace Transforms Related to Fractional Calculus

$\log \dfrac{s^2 + a^2}{s^2}$	$\dfrac{2}{t}(1 - \cos at)$
$\log \dfrac{s-a}{s-b}$	$\dfrac{1}{t}(e^{bt} - e^{at})$
$\dfrac{e^{-k\sqrt{s}}}{\sqrt{s}(a+\sqrt{s})},\ (k \geq 0)$	$e^{ak}e^{a^2 t}\,\mathrm{erfc}\!\left(a\sqrt{t}+\dfrac{k}{2\sqrt{t}}\right)$
$\dfrac{a e^{-k\sqrt{s}}}{s(a+\sqrt{s})},\ (k \geq 0)$	$\mathrm{erfc}\!\left(\dfrac{k}{2\sqrt{t}}\right) - e^{ak}e^{a^2 t}\,\mathrm{erfc}\!\left(a\sqrt{t}+\dfrac{k}{2\sqrt{t}}\right)$
$\dfrac{1}{s\sqrt{s}}e^{-k\sqrt{s}},\ (k \geq 0)$	$2\sqrt{\dfrac{t}{\pi}}e^{-\frac{k^2}{4t}} - k\cdot\mathrm{erfc}\!\left(\dfrac{k}{2\sqrt{t}}\right)$
$\dfrac{1}{\sqrt{s}}e^{-k\sqrt{s}},\ (k \geq 0)$	$\dfrac{1}{\sqrt{\pi.t}}e^{-\frac{k^2}{4t}}$
$\dfrac{1}{s}e^{-k\sqrt{s}},\ (k \geq 0)$	$\mathrm{erfc}\!\left(\dfrac{k}{2\sqrt{t}}\right)$
$e^{-k\sqrt{s}},\ (k \geq 0)$	$\dfrac{k}{2\sqrt{\pi.t^3}}e^{-\frac{k^2}{4t}}$
$\dfrac{1}{s^v}e^{k/s},\ (v > 0)$	$\left(\dfrac{t}{k}\right)^{(v-1)/2} I_{v-1}(2\sqrt{kt})$
$\dfrac{1}{s^v}e^{-k/s},\ (v > 0)$	$\left(\dfrac{t}{k}\right)^{(v-1)/2} J_{v-1}(2\sqrt{kt})$
$\dfrac{1}{s\sqrt{s}}e^{k/s}$	$\dfrac{1}{\sqrt{\pi.k}}\sinh 2\sqrt{kt}$
$\dfrac{1}{s\sqrt{s}}e^{-k/s}$	$\dfrac{1}{\sqrt{\pi.k}}\sin 2\sqrt{kt}$
$\dfrac{1}{\sqrt{s}}e^{k/s}$	$\dfrac{1}{\sqrt{\pi.t}}\cosh 2\sqrt{kt}$
$\dfrac{1}{\sqrt{s}}e^{-k/s}$	$\dfrac{1}{\sqrt{\pi.t}}\cos 2\sqrt{kt}$
$\dfrac{1}{s}e^{-k/s}$	$J_0(2\sqrt{kt})$
$\dfrac{k}{s^2+k^2}\coth\dfrac{\pi.s}{2k}$	$|\sin kt|$
$\dfrac{1}{\sqrt{s}}$	$\dfrac{1}{\sqrt{\pi.t}}$
$\dfrac{1}{s\sqrt{s}}$	$2\sqrt{\dfrac{t}{\pi}}$
$\dfrac{1}{s^n\sqrt{s}},\ (n = 1, 2, \ldots)$	$\dfrac{2^n t^{n-(1/2)}}{1.3.5\ldots(2n-1)\sqrt{\pi}}$
$\dfrac{s}{(s-a)^{3/2}}$	$\dfrac{1}{\sqrt{\pi.t}}e^{at}(1+2at)$
$\sqrt{s-a}-\sqrt{s-b}$	$\dfrac{1}{2\sqrt{\pi.t^3}}(e^{bt}-e^{at})$
$\dfrac{1}{\sqrt{s}+a}$	$\dfrac{1}{\sqrt{\pi.t}} - ae^{a^2 t}\,\mathrm{erfc}(a\sqrt{t})$
$\dfrac{\sqrt{s}}{s-a^2}$	$\dfrac{1}{\sqrt{\pi.t}} + ae^{a^2 t}\,\mathrm{erf}(a\sqrt{t})$

$\dfrac{\sqrt{s}}{s+a^2}$	$\dfrac{1}{\sqrt{\pi.t}} - \dfrac{2a}{\sqrt{\pi}} e^{-a^2 t} \int_0^{a\sqrt{t}} e^{\tau^2} d\tau$
$\dfrac{1}{\sqrt{s}(s-a^2)}$	$\dfrac{1}{a} e^{a^2 t} erf(a\sqrt{t})$
$\dfrac{1}{\sqrt{s}(s+a^2)}$	$\dfrac{2}{a\sqrt{\pi}} e^{-a^2 t} \int_0^{a\sqrt{t}} e^{\tau^2} d\tau$
$\dfrac{b^2 - a^2}{(s-a^2)(\sqrt{s}+b)}$	$e^{a^2 t}[b - a\{erf(a\sqrt{t})\}] - be^{b^2} erfc(b\sqrt{t})$
$\dfrac{1}{\sqrt{s}(\sqrt{s}+a)}$	$e^{a^2 t} erfc(a\sqrt{t})$
$\dfrac{1}{\sqrt{s+b}(s+a)}$	$\dfrac{1}{\sqrt{b-a}} e^{-at} erf(\sqrt{b-a}\sqrt{t})$
$\dfrac{b^2 - a^2}{\sqrt{s}(s-a^2)(\sqrt{s}+b)}$	$e^{a^2 t}\left[\dfrac{b}{a} erf(a\sqrt{t}) - 1\right] + e^{b^2 t} erfc(b\sqrt{t})$
$\dfrac{(1-s)^n}{s^{n+(1/2)}}$	$\dfrac{n!}{(2n)!\sqrt{\pi t}} H_{2n}(\sqrt{t})$
	$H_n(x) = e^{x^2} \dfrac{d^n}{dx^n}(e^{-x^2})$
	Hermetite polynomial
$\dfrac{(1-s)^n}{s^{n+(3/2)}}$	$-\dfrac{n!}{(2n+1)!\sqrt{\pi}} H_{2n+1}(\sqrt{t})$
$\dfrac{\sqrt{s+2a} - \sqrt{s}}{\sqrt{s}}$	$ae^{-at}[I_1(at) + I_0(at)]$
	$I_n(x) = j^{-n} J_n(jt)$
	J_n Bessel function of first kind
$\dfrac{1}{\sqrt{s+a}\sqrt{s+b}}$	$e^{-\frac{1}{2}(a+b)t} I_0\left(\dfrac{a-b}{2}t\right)$
$\dfrac{\Gamma(k)}{(s+a)^k(s+b)^k}$ For $k > 0$	$\sqrt{\pi}\left(\dfrac{t}{a-b}\right)^{k-(1/2)} e^{-\frac{1}{2}(a+b)t} I_{k-(1/2)}\left(\dfrac{a-b}{2}t\right)$
$\dfrac{1}{(s+a)^{1/2}(s+b)^{3/2}}$	$te^{-\frac{1}{2}(a+b)t}\left[I_0\left(\dfrac{a-b}{2}t\right) + I_1\left(\dfrac{a-b}{2}t\right)\right]$
$\dfrac{\sqrt{s+2a} - \sqrt{s}}{\sqrt{s+2a} + \sqrt{s}}$	$\dfrac{1}{t} e^{-at} I_1(at)$
$\dfrac{(a-b)^k}{(\sqrt{s+a}+\sqrt{s+b})^{2k}}$ For $k > 0$	$\dfrac{k}{t} e^{-\frac{1}{2}(a+b)t} I_k\left(\dfrac{a-b}{2}t\right)$
$\dfrac{1}{\sqrt{s}\sqrt{s+a}(\sqrt{s+a}+\sqrt{s})^{2v}}$ For $k > 0$	$\dfrac{1}{a^v} e^{-\frac{1}{2}at} I_v\left(\dfrac{a}{2}t\right)$
$\dfrac{1}{\sqrt{s^2+a^2}}$	$J_0(at)$
$\dfrac{1}{\sqrt{s^2-a^2}}$	$I_0(at)$, modified Bessel function of the first kind zero order

$\dfrac{(\sqrt{s^2+a^2}-s)^v}{\sqrt{s^2+a^2}}\quad (v>-1)$	$a^v J_v(at)$
$\dfrac{1}{(\sqrt{s^2+a^2})^k}\quad k>0$	$\dfrac{\sqrt{\pi}}{\Gamma(k)}\left(\dfrac{t}{2a}\right)^{k-(1/2)} J_{k-(1/2)}(at)$
$(\sqrt{s^2+a^2}-s)^k,\quad (k>0)$	$\dfrac{ka^k}{t} J_k(at)$
$\dfrac{(\sqrt{s^2-a^2}+s)^v}{\sqrt{s^2-a^2}},\quad (v>-1)$	$a^v I_v(at)$
$\dfrac{1}{(s^2-a^2)^k},(k>0)$	$\dfrac{\sqrt{\pi}}{\Gamma(k)}\left(\dfrac{t}{2a}\right)^{k-(1/2)} I_{k-(1/2)}(at)$
$\dfrac{1}{s\sqrt{s+1}}$	$erf(\sqrt{t})$
$\dfrac{1}{s+\sqrt{s^2+a^2}}$	$\dfrac{J_1(at)}{at}$
$\dfrac{1}{(s+\sqrt{s^2+a^2})^N}$	$\dfrac{NJ_N(at)}{a^N t}$
$\dfrac{1}{\sqrt{s^2+a^2}(s+\sqrt{s^2+a^2})}$	$\dfrac{J_1(at)}{a}$
$\dfrac{1}{\sqrt{s^2+a^2}(s+\sqrt{s^2+a^2})^N}$	$\dfrac{J_N(at)}{a^N}$

2.4 Concluding Comments

In this chapter, the basis functions that are important in the study of the fractional order systems are introduced. Mostly the fundamental form is the Mittag-Leffler function, and thus can be stated as the generalized exponential function. As the exponential function plays the basis role in integer order calculus, so does Mittag-Leffler function has its role in the fractional calculus. Other compacted forms of the variants of Mittag-Leffler variety are also listed, which find several applications of solution of fractional differential equations. All of these functions are of power-series expansions and fit a variety of power law following processes. In conclusion, the readers will be put to think about following reality. In circuit theory experiment, we have made a low pass filter, with lumped resistance and lumped capacitor. The step response to this should have a pure exponential reaction, and mostly the recorders will show the similar reaction. The question is are we observing a pure exponential curve uniquely determined by unique time constant, the product of lumped resistance and lumped capacitor used. We tend to believe that the observation is pure exponential and ode the aberration to non-linearity, instrument error, leakages, distributed effects, and various others like parametric drifts of components. The deviation from the expected curve, if redrawn by a suitable 'power series' function of Mittag-Leffler type, then we question the descriptor equation, which classically is an integer order differential equation with 'lumped circuit components'. The Mittag-Leffler type function is the solution of fractional differential equation, thus if the basic circuit descriptor were of fractional order differential equation then we explain the reality, closely. However no capacitor is pure capacitor, no resistance is pure resistance, and no system can have lumped characteristic, and the distributed parametric spread is reality. The same thoughts can be extended to various other relaxation processes of the nature about diffusion, reactor kinetics, electrochemistry, and several others. The variants of Mittag-Leffler functions introduced here are developed in last four decades; several others may be developed in future to explain the physical processes of nature.

Chapter 3
Observation of Fractional Calculus in Physical System Description

3.1 Introduction

Fractional calculus allows a more compact representation and problem solution for some spatially distributed systems. Spatially distributed system representation allows a better understanding of the fractional calculus. The idea of fractional integrals and derivatives has been known since the development of regular calculus. Although not well known to most engineers, prominent mathematicians as well as scientists of the operational calculus have considered the fractional calculus. Unfortunately, many of the results in the fractional calculus are given in the language of advanced analysis and are not readily accessible to the general engineering and science community. Many systems are known to display fractional order dynamics. Probably the first physical system to be widely recognized as one demonstrating fractional behavior is the semi-infinite lossy (RC) transmission line. The current into the line is equal to the half derivative of the applied voltage. That is, impedance is

$$V(s) = \frac{1}{\sqrt{s}} I(s) \ ;$$

many studied this system, Heaviside (1871) considered it extensively using the operational calculus. He states that "there is universe of mathematics lying in between the complete differentiations and integrations, and that fractional operators push themselves forward sometimes, and are just as real as others." Another equivalent system is diffusion of heat into semi-infinite solid. Here temperature looking in from the boundary is equal to the half integral of the heat rate there. Other systems that are known to display fractional order dynamics are viscoelasticity, colored noise, electrode–electrolyte polarization, dielectric polarization, boundary layer effects in ducts, and electromagnetic waves. Because many of these systems depend upon specific material and chemical properties, it is expected that a wide range of fractional order behaviors are possible using different materials.

3.2 Temperature–Heat Flux Relationship for Heat Flowing in Semi-infinite Conductor

The thermocouple consists of two pairs of dissimilar metals with a common junction point. Because the wires are long and insulated, they will be treated as "semi-infinite" heat conductors. Figure 3.1 represents one such wire of thermocouple

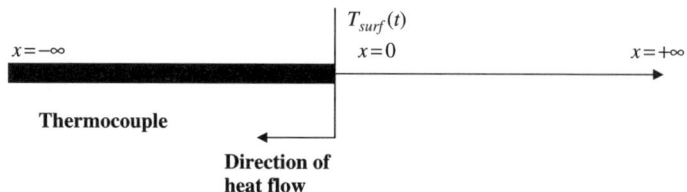

Fig. 3.1 Heat flow in semi-infinite wire thermocouple

pair. The thick line in Fig. 3.1 represents the semi-infinite heat conductor, the thermocouple wire measuring the temperature at $x = 0$ the furnace wall, called as $T_{surf}(t)$, which dynamically varies with the time. The initial temperature is denoted by T_0

The problem of heat conduction in the thermocouple wire is obviously one-dimensional. The following derivation shows how fractional calculus appears in the problem of relating the conduction heat flux through semi-infinite thermocouple wire to the body temperature at the origin.

$$c\rho \frac{\partial T}{\partial t} = k \frac{\partial^2 T}{\partial x^2},$$
$$(t > 0, \, \& -\infty < x < 0)$$
$$T(0, x) = T_0$$
$$T(t, 0) = T_{surf}(t)$$
$$\left| \lim_{x \to -\infty} T(t, x) \right| < \infty$$

where t is time(s), x is the spatial direction in the direction of heat flow (m), c is the specific heat or heat capacity (J kg^{-1} K^{-1}), ρ is density (kg m^{-3}), $T(t, x)$ is the temperature (K), and k is coefficient of heat conduction (W m^{-1} K^{-1}).

Let $u(t, x) = T(t, x) - T_0$. Substituting this in the above set of equations, we get

$$c\rho \frac{\partial u}{\partial t} = k \frac{\partial^2 u}{\partial x^2},$$
$$(t > 0, \, \& -\infty < x < 0)$$
$$u(0, x) = 0$$
$$u(t, 0) = T_{surf}(t) - T_0$$
$$\left| \lim_{x \to -\infty} u(t, x) \right| < \infty$$

3.2 Temperature–Heat Flux Relationship for Heat Flowing in Semi-infinite Conductor

Taking Laplace transforms for the above equation gives

$$c\rho.sU(s,x) = k\frac{\partial^2 U(s,x)}{\partial x^2}$$

$$\frac{\partial^2 U(s,x)}{\partial x^2} - \frac{c\rho s}{k}U(s,x) = 0$$

The bounded solution for x tends to $-\infty$ is

$$U(s,x) = U(s,0)\exp\left(x\sqrt{\frac{sc\rho}{k}}\right);$$

differentiating this, we find

$$\frac{dU(s,x)}{dx} = U(s,0)\sqrt{\frac{sc\rho}{k}}\exp\left(x\sqrt{\frac{sc\rho}{k}}\right).$$

From these two expressions, we get the following by putting $x = 0$ and taking the inverse Laplace of $s^{-0.5}F(s) \to d^{-1/2}f(t)$, i.e., semi-integration, we obtain semi-differential equation in time variable:

$$\frac{1}{\sqrt{s}}\frac{d}{dx}U(s,0) = \sqrt{\frac{c\rho}{k}}U(s,0)$$

$$\frac{d^{-1/2}}{dt^{-1/2}}\frac{\partial u(t,0)}{\partial x} = \sqrt{\frac{c\rho}{k}}u(t,0)$$

$$\frac{\partial u(t,0)}{\partial x} = \sqrt{\frac{c\rho}{k}}\frac{d^{1/2}}{dt^{1/2}}u(t,0)$$

Returning from $u(t,x)$ to $T(t,x)$, we get

$$k\frac{\partial T(t,0)}{\partial x} = \sqrt{c\rho k}\frac{d^{1/2}}{dt^{1/2}}(T_{surf}\{t\} - T_0).$$

$$k\frac{\partial}{\partial x}T(t,0) = Q(t)$$

is termed as heat flux, flowing through the thermocouple wire at the interface of the furnace wall and point of contact (the origin). Therefore, the heat flux expression is

$$Q(t) = \sqrt{c\rho k}\frac{d^{1/2}}{dt^{1/2}}(T_{surf}(t) - T_0) = \frac{k}{\sqrt{\frac{k}{c\rho}}}\frac{d^{1/2}}{dt^{1/2}}(T_{surf}(t) - T_0)$$

$$= \frac{k}{\sqrt{\alpha}}{}_aD_t^{1/2}T_b(t).$$

3.3 Single Thermocouple Junction Temperature in Measurement of Heat Flux

From the derivation, as in the above section, we can write a general heat flow equation relating the heat flux conducted through a semi-infinite conductor of heat to the temperature at the origin as time varying constitutive relation:

$$Q_i(t) = \frac{k}{\sqrt{\alpha}} {}_aD_t^{1/2} T_b,$$

$${}_aD_t^{1/2} \equiv \frac{d^{1/2}}{[d(t-a)]^{1/2}}$$

$$\alpha = \frac{k}{c\rho}$$

The semi-derivative is shown for initial time point a. When initial forcing conditions (states) are zero, then the operator is

$${}_0D_t^{1/2} \equiv \frac{d^{1/2}}{dt^{1/2}} \ ;$$

here α is thermal diffusivity; T_b is the body temperature, at the point of contact of thermocouple to the furnace wall.

The following equations define the time domain behavior.

Input heat flux to the thermocouple from steam temperature to the tip of the thermocouple junction $Q_i = hA(T_g(t) - T_b(t))$. At the tip of the thermocouple, this input heat flux flows into two thermocouple wires as shown in Fig. 3.2. Thus,

$$Q_i(t) - Q_1(t) - Q_2(t) = mc\frac{dT_b}{dt} \ .$$

Converting this expression to integral form, we obtain the thermocouple node temperature related to two heat fluxes as

$$T_b(t) = \frac{1}{mc} {}_aD_t^{-1}(Q_i(t) - Q_1(t) - Q_2(t)) \ .$$

The two semi-infinite heat conductors have constitutive equations in semi-differential form as derived for Fig. 3.1, as

Fig. 3.2 Thermocouple junction for temperature (heat flux) measurement

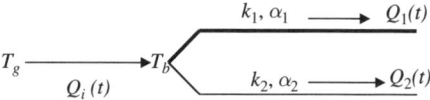

3.3 Single Thermocouple Junction Temperature in Measurement of Heat Flux

$$Q_1(t) = \frac{k_1}{\sqrt{\alpha_1}} {}_aD_t^{1/2} T_b(t),$$

and

$$Q_2(t) = \frac{k_2}{\sqrt{\alpha_2}} {}_aD_t^{1/2} T_b(t)$$

where hA is product of convective heat transfer coefficient and surface area, and mc is product of the mass and specific heat. The constitutive equation is obtained by substituting values of $Q's$, as

$$hA\left(T_g(t) - T_b(t)\right) - \frac{k_1}{\sqrt{\alpha_1}} {}_aD_t^{1/2} T_b(t) - \frac{k_1}{\sqrt{\alpha_2}} {}_aD_t^{1/2} T_b(t) = mc \frac{dT_b(t)}{dt},$$

after taking Laplace transforms of the constitutive equations, we have the following expression:

$$\left(mcs + \frac{k_1}{\sqrt{\alpha_1}} s^{1/2} + \frac{k_2}{\sqrt{\alpha_2}} s^{1/2} + hA\right) T_b(s) = hAT_g(s).$$

The transfer function is as follows:

$$\frac{T_b(s)}{T_g(s)} = \frac{1}{\left(\frac{mc}{hA}\right)s + \frac{1}{hA}\left(\frac{k_1}{\sqrt{\alpha_1}} + \frac{k_2}{\sqrt{\alpha_2}}\right)s^{1/2} + 1}$$

The value of fractional calculus is clearly demonstrated in this analysis. Conventional approaches require the solution of two simultaneous partial differential equations with ordinary integer order differential equation. The Bode plots show two distinct asymptotes: (1) slope -10 db/decade (corresponds to semi-pole $s^{1/2}$ behavior) and (2) as -20 db/decade at higher frequency. The diagram is shown in Fig. 3.3 with $(mc/hA) = 0.005$, and

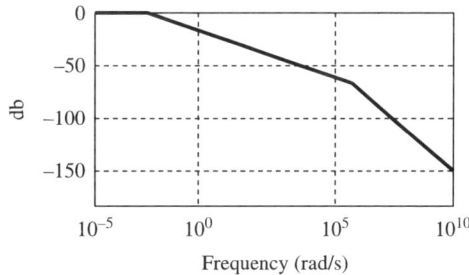

Fig. 3.3 Frequency response amplitude frequency Bode plot

$$\frac{1}{hA}\left(\frac{k_1}{\sqrt{\alpha_1}} + \frac{k_2}{\sqrt{\alpha_2}}\right) = 5.0.$$

One more observation may be drawn from this analysis. In order to estimate heat flux in any thermal system, classical method of utilizing two thermocouples can be replaced by one thermocouple and from the temperature values obtained as function of time instantaneous semi-differential equation, as indicated above, may be solved to estimate flowing heat flux.

Observing the transfer function of this system, $T_b(s)/T_g(s)$, obtained earlier, it appears as per integer order calculus theory that it is a first-order system. The first-order system response to a step input is a damped output, without any oscillation and overshoot. The presence of fractional order terms makes somewhat an anomalous argument (which is detailed in Chap. 9.9). Though the system appears to be of first order, yet the presence of the half-order term, in denominator, may give a system response, to a step input as oscillatory with overshoot. Therefore definition of the system order for fractional order system is different, than that in integer order calculus.

3.4 Heat Transfer

System identification is the part of control practice, in which the parameters that enter into mathematical model of the system are determined. This is especially important in thermal system with convection because of presence of heat transfer coefficient, which is never really known with great exactitude, and may vary with time due to physical or chemical changes at the heat transfer surface. There are also other parameters like the thermal capacity of the conductive body, its surface area, and its thermal diffusivity that have to be determined.

Rather than to find each of the parameters separately, it is more practical to estimate non-dimensional expression that best fits the observed data. In this heat transfer example, consider the cooling (or heating) of a one-dimensional plane wall of thickness L, with spatial uniform initial temperature T_i. There is convective heat transfer coefficient h from one wall to fluid at temperature T_∞. The exact solution with partial differential equation is as follows with the boundary and initial conditions.

The temperature field is $T(x, t)$, where x is the coordinate measured from the wall, and t is the time. The transient heat conduction equation in the wall is given by

$$\frac{\partial T(x,t)}{\partial t} = \alpha \frac{\partial^2 T(x,t)}{\partial x^2},$$

where α is the thermal diffusivity.

3.4 Heat Transfer

The initial and boundary conditions are

$$\frac{\partial T(x,t)}{\partial x} = 0 \text{ at } x = 0$$

$$k\frac{\partial T(x,t)}{\partial x} + h(T\{x,t\} - T_\infty) = 0, \text{ at } x = L$$

$$T(x,t) = T_i, \text{ at } t = 0$$

where k is the thermal conductivity of the wall material. With change of variable to make dimensionless equation, we get the following transformed unitless variables as

$$\xi = \frac{x}{L},$$

$$\tau = \frac{t\alpha}{L^2},$$

and unitless temperature as

$$\theta = \frac{T(x,t) - T_\infty}{T_i - T_\infty},$$

and we obtain the dimensionless equation as

$$\frac{\partial \theta}{\partial \tau} = \frac{\partial^2 \theta}{\partial \xi^2},$$

with $\frac{\partial \theta}{\partial \xi} = 0$, at $\xi = 0$ and $\frac{\partial \theta}{\partial \xi} + B_i\theta = 0$ at $\xi = 1$, and for $\tau = 0$.

Where the "size-factor" $B_i = hL/k$ is called the Biot's number.

Solution to this dimensionless equation, which is the exact representation of heat transfer, is

$$\theta(x,\tau) = \sum_{n=1}^{\infty} C_n \exp\left(-\lambda_n^2 \tau\right) \cos(\lambda_n x),$$

where

$$C_n = \frac{4\sin\lambda_n}{2\lambda_n + \sin(2\lambda_n)}$$

and λ_n are positive roots of transcendental expression $\lambda_n \tan \lambda_n = B_i$.

The dimensionless mean temperature is

$$\bar{\theta}(\tau) = \int_0^1 \theta(\xi, \tau) d\xi.$$

The exact solution is obtained above, by considering various Biot's number (0.1–10). However, the first way to approximate this heat transfer phenomena is by having spatial average of the temperature, i.e.

$$\bar{T}(t) = \frac{1}{L} \int_0^L T(x,t)dx$$

be taken as dependent variable. In terms of this average temperature, the heat balance equation is

$$\frac{d\bar{T}(t)}{dt} + \frac{h}{\rho c}\left(\bar{T} - T_\infty\right) = 0,$$

where ρ is the density and c is specific heat of the wall material. The convective heat transfer initial and boundary conditions $\partial T/\partial x = 0$, at $x = 0$, and $k(\partial T/\partial x) + h(T - T_\infty) = 0$, at $x = 1$ have been used to derive the above average expression as approximation. The equation $T = T_i$ at $t = 0$ gives $\bar{T}(0) = 1$. Using dimensionless variables, as done for exact solution case, one obtains $d\bar{\theta}/d\tau + B_i \bar{\theta} = 0$, with $\bar{\theta}(0) = 1$. Now numerical experiments point toward an interesting observation that the solution with $B_i = 0.1$, the exact solution and this approximation match closely, whereas for $B_i = 10$, the deviations are large.

An improvement to the above approximation is to write a fractional order differential equation, as $(d^q \bar{\theta}/d\tau^q) + pB_i \bar{\theta} = 0$, with $\bar{\theta}(0) = 1$. Here q and p be varied to minimize

$$E = \int_0^{\tau_{max}} e(\tau)^2 d\tau,$$

where $e(\tau)$ is the difference between the exact solution and approximate solution of $\bar{\theta}(\tau)$. τ_{max} is the maximum value of τ to which integration is carried out.

Here the effect of B_i is to be discussed. The fractional order $q \to 1$ and the multiplier of the Biot's number $p \to 1$, as $B_i \to 0$.

This example works well with dynamic system modeling where measurements enable simultaneous time-dependent system identification as well as provide an error signal for feedback controls. In the present heat transfer example, for the wall with simple geometry, a lumped parameter energy balance in which temperature of the system is assumed to be spatially uniform is commonly used to model transient conductive systems exposed to convective heat fluxes at their boundaries. It is simple to fit experimental data and the integer order (in this case first order) differential equation that is easy to solve. For larger size (Biot's number), there is difference between actual temperature field and the spatial average used in lumped model. This necessitates the solution of the partial differential equations for the transient heat conducting in the body which is made difficult by shape, heterogeneity, or the unknown nature of the convective heat transfer coefficient at the boundaries.

This argument is true for any realistic process with transport phenomena. Here a practical method to describe the system with fractional order differential equation aims at reality in system identification and control, as real systems are distributed phenomena.

3.5 Driving Point Impedance of Semi-infinite Lossy Transmission Line

Assuming a lossy RC line, the boundary value problem can be defined in terms of the current or voltage variables. Since a semi-infinite line is considered, the measurable inputs or outputs are at $x = 0$, at the left end, while the right end $x = \infty$ is at finite value. In terms of the voltage variable, the equations can be written as

$$\frac{\partial v(x,t)}{\partial x} = i(x,t)R$$
$$\frac{\partial i(x,t)}{\partial x} = C\frac{\partial v(x,t)}{\partial t}$$

R and C are resistance and capacitance per unit length.

Differentiating first with respect to x and then substituting second in the first one, we get

$$\frac{\partial^2 v}{\partial x^2} = R\frac{\partial i}{\partial x} = RC\frac{\partial v}{\partial t},$$

choosing $1/RC$ as α, we get the problem formulation as

$$\frac{\partial v(x,t)}{\partial t} = \alpha\frac{\partial^2 v(x,t)}{\partial x^2},\ v(0.t) = v_I(t),\ v(\infty,t) = 0.v(x,0)$$

Given with

$$i(x,t) = -\frac{1}{R}\frac{\partial v(x,t)}{\partial x}$$

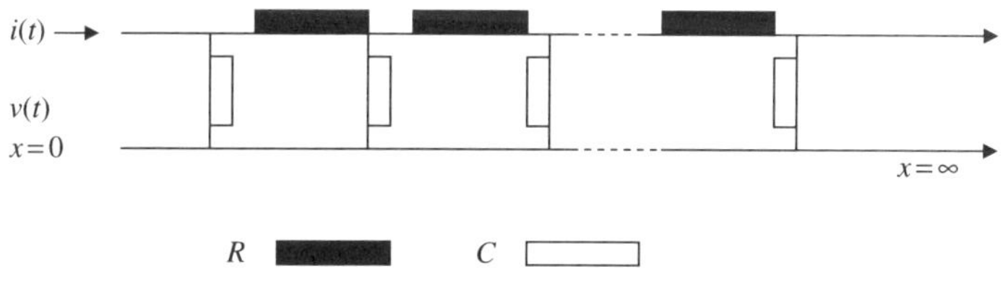

Fig. 3.4 Semi-infinite lossy transmission line

In this formulation v is the voltage, i is the current, and $v_I(t)$ is a time-dependent input variable. At $x = \infty$, the condition is of short circuit. A classical solution using iterated Laplace transform is used to solve this problem.

Taking Laplace transform with respect to time and using s as temporal Laplace variable gives

$$sV(x,s) - v(x,0) = \alpha \frac{d^2 V(x,s)}{dx^2}, \text{ with } V(0,s) = V_I(s), V(\infty, s) = 0,$$

$$I(x,s) = -\frac{1}{R}\frac{dV(x,s)}{dx}.$$

Then taking the Laplace transform with respect to spatial position x, and using p as the spatial Laplace variable gives

$$\frac{s}{\alpha}V(p,s) - \frac{1}{\alpha}V(p,0) = p^2 V(p,s) - pV(0,s) - \left[\frac{dV(0,s)}{dx}\right].$$

Substituting

$$\left[\frac{dV(0,s)}{dx}\right] = V^*,$$

this equation can be manipulated to give $V(p,s)$ as

$$\left[p^2 - \frac{s}{\alpha}\right]V(p,s) = -\frac{1}{\alpha}V(p,0) + pV(0,s) + V^*(0,s),$$

or

$$V(p,s) = \left[\frac{1}{p^2 - \frac{s}{\alpha}}\right]\left[-\frac{1}{\alpha}V(p,0) + pV(0,s) + V^*(0,s)\right]$$

The first term of the transform, here, of the initial spatial distribution $V(p,0)$ is problem-dependent. After rearrangement and partial fraction, the above expression can be expressed as

$$V(p,s) = \left[\frac{1}{2\sqrt{\frac{s}{\alpha}}\left(p - \sqrt{\frac{s}{\alpha}}\right)} - \frac{1}{2\sqrt{\frac{s}{\alpha}}\left(p - \sqrt{\frac{s}{\alpha}}\right)}\right]\left[-\frac{1}{\alpha}V(p,0)\right]$$
$$+ \left[\frac{1}{p^2 - \frac{s}{\alpha}}\right]\left[pV(0,s) + V^*(0,s)\right]$$

Here the first term represents the effect of any initial spatial voltage distribution, while the second term represents the voltage and current present at $x = 0$ at the end of the line. The first term is now inverse Laplace transformed with respect to the variable p using convolution, and the second is inverse-Laplace transformed using

3.5 Driving Point Impedance of Semi-infinite Lossy Transmission Line

standard transform pairs.

$$V(x,s) = \int_0^x \frac{1}{2\sqrt{\frac{s}{\alpha}}} e^{+(x-\lambda)\sqrt{\frac{s}{\alpha}}} \left[-\frac{1}{\alpha}v(\lambda,0)\right] d\lambda$$

$$-\int_0^x \frac{1}{2\sqrt{\frac{s}{\alpha}}} e^{-(x-\lambda)\sqrt{\frac{s}{\alpha}}} \left[-\frac{1}{\alpha}v(\lambda,0)\right] d\lambda$$

$$+V(0,s)\cosh\left(x\sqrt{\frac{s}{\alpha}}\right) + \frac{V^*(0,s)}{\sqrt{\frac{s}{\alpha}}}\sinh\left(x\sqrt{\frac{s}{\alpha}}\right)$$

Equivalently,

$$V(x,s) = \int_0^x \frac{1}{2\sqrt{\frac{s}{\alpha}}} e^{+x\sqrt{\frac{s}{\alpha}}} e^{-\lambda\sqrt{\frac{s}{\alpha}}} \left[-\frac{1}{\alpha}v(\lambda,0)\right] d\lambda$$

$$-\int_0^x \frac{1}{2\sqrt{\frac{s}{\alpha}}} e^{-x\sqrt{\frac{s}{\alpha}}} e^{+\lambda\sqrt{\frac{s}{\alpha}}} \left[-\frac{1}{\alpha}v(\lambda,0)\right] d\lambda$$

$$+\frac{V(0,s)}{2}\left[e^{+x\sqrt{\frac{s}{\alpha}}} + e^{-x\sqrt{\frac{s}{\alpha}}}\right] + \frac{V^*(0,s)}{2\sqrt{\frac{s}{\alpha}}}\left[e^{+x\sqrt{\frac{s}{\alpha}}} - e^{-x\sqrt{\frac{s}{\alpha}}}\right]$$

Collecting the like exponentials gives the following:

$$V(x,s) = \frac{e^{+x\sqrt{\frac{s}{\alpha}}}}{2}\left[V(0,s) + \frac{V^*(0,s)}{\sqrt{\frac{s}{\alpha}}} - \frac{1}{\alpha\sqrt{\frac{s}{\alpha}}}\int_0^x e^{-\lambda\sqrt{\frac{s}{\alpha}}} v(\lambda,0) d\lambda\right]$$

$$+\frac{e^{-x\sqrt{\frac{s}{\alpha}}}}{2}\left[V(0,s) - \frac{V^*(0,s)}{\sqrt{\frac{s}{\alpha}}} + \frac{1}{\alpha\sqrt{\frac{s}{\alpha}}}\int_0^x e^{+\lambda\sqrt{\frac{s}{\alpha}}} v(\lambda,0) d\lambda\right]$$

It should be recognized that the coefficients multiplying the two exponential functions are unknowns. Although the integral and either $V(0,s)$ or $V^*(0,s)$ are given in the problem statement, the other condition (V^* or V, respectively) at $x = 0$ is determined as a response to these two given terms. Imposing the boundary condition at $x = \infty$, allows the determination of a relationship between these three terms at $x = 0$, and thus allows the impedance and initial condition response of the system.

It is required to evaluate the above equation in the limit $x \to \infty$. In this limit second term in the above equation goes to zero due to exponential behavior, however the integral inside the bracket will diverge. That is

$$\lim_{x \to \infty} \left(e^{-x\sqrt{\frac{s}{\alpha}}}\right) \left(\frac{1}{\alpha\sqrt{\frac{s}{\alpha}}} \int_0^x e^{+\lambda\sqrt{\frac{s}{\alpha}}} v(\lambda,0) d\lambda\right) = 0.\infty$$

We are thus left with indeterminate form, and this can be solved by L'Hopital's rule as follows (after rearrangement):

$$\lim_{x \to \infty} \frac{\frac{1}{2\alpha\sqrt{\frac{s}{\alpha}}} \int_0^x e^{+\lambda\sqrt{\frac{s}{\alpha}}} v(\lambda, 0) d\lambda}{e^{+x\sqrt{\frac{s}{\alpha}}}}$$

The L'Hopital rule says that this ratio has the same value as the ratio of the derivatives (with respect to x) of the numerator and denominator. Differentiating the denominator is easy, but differentiating the numerator with respect to x requires Leibniz's rule. Performing the differentiation gives

$$\lim_{x \to \infty} \frac{\left(\frac{1}{2\alpha\sqrt{\frac{s}{\alpha}}} \frac{d}{dx} \int_0^x e^{+\lambda\sqrt{\frac{s}{\alpha}}} v(\lambda, 0) d\lambda \right)}{\left(\sqrt{\frac{s}{\alpha}} e^{+x\sqrt{\frac{s}{\alpha}}} \right)}$$

Combining the leading constants, and applying the Leibniz's rule to the numerator gives

$$\lim_{x \to \infty} \frac{\frac{1}{2s} e^{+x\sqrt{\frac{s}{\alpha}}} v(x, 0)}{e^{+x\sqrt{\frac{s}{\alpha}}}}.$$

It can now be seen that the exponential terms cancel, which leaves the result

$$\lim_{x \to \infty} \frac{1}{2s} v(x, 0) = \frac{v(\infty, 0)}{2s}.$$

The problem statement however requires that the boundary condition $v(\infty, t) = 0$ be satisfied for all time. Thus it is shown that the first term of the main equation equals zero for $x \to \infty$.

From above limit derivations,

$$V(\infty, s) = \frac{e^{+\infty\sqrt{\frac{s}{\alpha}}}}{2} \left[V(0, s) + \frac{V^*(0, s)}{\sqrt{\frac{s}{\alpha}}} - \lim_{x \to \infty} \frac{1}{\alpha\sqrt{\frac{s}{\alpha}}} \int_0^x e^{-\lambda\sqrt{\frac{s}{\alpha}}} v(\lambda, 0) d\lambda \right] = 0$$

dropping the limit notation, we have

$$V(0, s) + \frac{V^*(0, s)}{\sqrt{\frac{s}{\alpha}}} - \frac{1}{\alpha\sqrt{\frac{s}{\alpha}}} \int_0^\infty e^{-\lambda\sqrt{\frac{s}{\alpha}}} v(\lambda, 0) d\lambda = 0.$$

Remembering that the current anywhere in the line is related to the voltage, then at $x = 0$

$$I(0, s) = -\frac{1}{R} \frac{dV(0, s)}{dx} = -\frac{V^*(0, s)}{R}$$

3.5 Driving Point Impedance of Semi-infinite Lossy Transmission Line

and solving for voltage in terms of source current gives

$$V(0,s) = \frac{RI(0,s)}{\sqrt{\frac{s}{\alpha}}} + \frac{1}{\alpha\sqrt{\frac{s}{\alpha}}} \int_0^\infty e^{-\lambda\sqrt{\frac{s}{\alpha}}} v(\lambda, 0) d\lambda$$

In evaluating the integral on the right, it is now recognized that this term is equivalent to a Laplace transform integral with

$$s \to q = \sqrt{\frac{s}{\alpha}}.$$

Thus the Laplace transform tables can simplify the evaluation of this term as follows:

$$\frac{1}{\alpha\sqrt{\frac{s}{\alpha}}} \int_0^\infty e^{-\lambda\sqrt{\frac{s}{\alpha}}} v(\lambda, 0) d\lambda = \frac{1}{\alpha\sqrt{\frac{s}{\alpha}}} [V(q,0)]_{q=\sqrt{\frac{s}{\alpha}}}.$$

The notation here on the right-hand side of this equation is used to indicate the evaluation procedure. First the initial spatial distribution $v(x, 0)$ is Laplace transformed with respect to the spatial Laplace variable p to give $V(p, 0)$. The integral on the left side of the above equation is then easily calculated by replacing spatial variable p with

$$q = \sqrt{\frac{s}{\alpha}}.$$

The voltage equation thus becomes

$$V(0,s) = \frac{RI(0,s)}{\sqrt{\frac{s}{\alpha}}} + \frac{1}{\alpha\sqrt{\frac{s}{\alpha}}} [V(p,0)]_{p=\sqrt{\frac{s}{\alpha}}}.$$

Notice that this contains the driving point impedance function $Z(s)$, which is obtained by setting the initial condition, which terms to zero.

$$Z(s) = \frac{V(0,s)}{I(0,s)} = \frac{R}{\sqrt{\frac{s}{\alpha}}}$$

or as $\alpha = 1/RC$, the impedance is

$$Z(s) = \frac{V(0,s)}{I(0,s)} = \sqrt{\frac{R}{C}} \frac{1}{\sqrt{s}}.$$

Note that in the impedance expression of $Z(s)$, there are two parts, the forced response due to $I(0, s)$ and the initial condition response due to the initial voltage distribution in the lossy line. The final expression of voltage anywhere in the line as function of the applied voltage at the terminal $V_I(s)$ and the initial condition on the line is

$$V(x, s) = \frac{e^{+x\sqrt{\frac{s}{\alpha}}}}{2\alpha\sqrt{\frac{s}{\alpha}}} \left[[V(p, 0)]_{p=\sqrt{\frac{s}{\alpha}}} - \int_0^x e^{-\lambda\sqrt{\frac{s}{\alpha}}} v(\lambda, 0) d\lambda \right]$$

$$+ \frac{e^{-x\sqrt{\frac{s}{\alpha}}}}{2\alpha\sqrt{\frac{s}{\alpha}}} \left[2V_I(s) - \frac{[V(p, 0)]_{p=\sqrt{\frac{s}{\alpha}}}}{\alpha\sqrt{\frac{s}{\alpha}}} + \frac{1}{\alpha\sqrt{\frac{s}{\alpha}}} \int_0^x e^{+\lambda\sqrt{\frac{s}{\alpha}}} v(\lambda, 0) d\lambda \right]$$

Furthermore, the current at any point in the line can be determined directly from the above equation as

$$I(x, s) = -\frac{1}{R} \frac{dV(x, s)}{dx}$$

By applying inverse Laplace transforms, for the driving point impedance expression, the voltage–current behavior (with zero initial condition) is obtained as

$$i(t) = \frac{1}{R\sqrt{\alpha}} \frac{d^{1/2}v(t)}{dt^{1/2}}$$

$$v(t) = R\sqrt{\alpha} \frac{d^{-1/2}v(t)}{dt^{-1/2}}$$

More compactly the voltage–current relation with the initial condition is expressed as

$$v(t) = R\sqrt{\alpha} \frac{d^{-1/2}i(t)}{dt^{-1/2}} + \varphi_1(t),$$

$$\varphi_1(t) = L^{-1} \left[\frac{1}{\alpha\sqrt{\frac{s}{\alpha}}} [V(p, 0)]_{p=\sqrt{\frac{s}{\alpha}}} \right] = L^{-1} \left(\frac{1}{\alpha\sqrt{\frac{s}{\alpha}}} \int_0^x e^{-\lambda\sqrt{\frac{s}{\alpha}}} v(\lambda, 0) d\lambda \right)$$

3.5 Driving Point Impedance of Semi-infinite Lossy Transmission Line

or

$$i(t) = \frac{1}{R\sqrt{\alpha}} \frac{d^{1/2}v(t)}{dt^{1/2}} + \varphi_2(t)$$

$$\varphi_2(t) = \frac{1}{R\sqrt{\alpha}} \frac{d\varphi_1(t)}{dt}$$

3.5.1 Practical Application of the Semi-infinite Line in Circuits

3.5.1.1 Semi-integrator Circuit

The circuit shown in the Fig. 3.5 performs the function of semi- integration of the input voltage $v_i(t)$. The half-order element (semi-infinite lossy line) is based on one-dimensional diffusion equation

$$\frac{\partial v}{\partial t} = \alpha \frac{\partial^2 v}{\partial x^2},$$

which is depicted by a ladder of discrete resistance and capacitance as shown in Fig. 3.4, and its connection is shown in Fig. 3.5 in an operational amplifier circuit. The terminal characteristic or the driving point impedance as obtained is described as

$$v(t) = r\sqrt{\alpha} \frac{d^{-1/2}i(t)}{dt^{-1/2}} + \varphi_1(t)$$

or

$$i(t) = \frac{1}{r\sqrt{\alpha}} \frac{d^{1/2}v(t)}{dt^{1/2}} + \varphi_2(t).$$

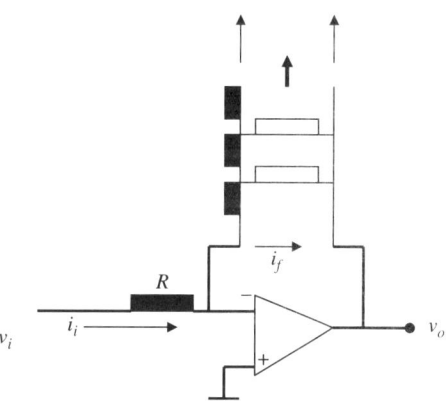

Fig. 3.5 Semi-integrator

Here $v(t)$ and $i(t)$ are the voltage and current respectively, at the terminal element, r is the resistance per unit length and α is the product of r and c (the capacitance per unit length of the line). The initial condition functions are determined by the initial state of charge and voltage or current that exists on the infinite array of elements. For operational amplifier, negative feedback configuration is

$$v_i(t) - 0 = i_i(t)R$$
$$0 - v_0(t) = r\sqrt{\alpha}\frac{d^{-1/2}i_f(t)}{dt^{-1/2}} + \varphi_I(t) = r\sqrt{\alpha}_c D_t^{-1/2} i_f(t)$$
$$i_i(t) = i_f(t)$$

solving for $v_o(t)$

$$v_0(t) = -r\sqrt{\alpha}_c D_t^{-1/2}\left\{\frac{1}{R}v_i(t)\right\}$$
$$v_0(t) = -\frac{r\sqrt{\alpha}}{R}{}_c D_t^{-1/2} v_i(t)$$

Note the symbolism change from smallcase differential operator to uppercase one, where the initialization function got included. This will be taken in detail while elaborate explanation of the initialization of fractional differintegrals is given in Chap. 6.

This is the basis of semi-integrator computing element. The equivalent (un-initialized) impedance form may also be calculated as $Z_f = r\sqrt{\alpha}/s^{1/2}$, $Z_i = R$. The transfer function (un-initialized) form is thus is

$$\frac{v_0(s)}{v_i(s)} = -\frac{r\sqrt{\alpha}}{Rs^{1/2}}$$

3.5.1.2 Semi-differentiator Circuit

For the circuit in Fig. 3.6, the negative feedback configuration gives

$$i_i(t) = \frac{1}{r\sqrt{\alpha}}{}_c D_t^{1/2}(v_i(t) - 0)$$
$$0 - v_o(t) = Ri_f(t)$$
$$i_i(t) = i_f(t)$$
$$v_o(t) = -Ri_f(t) = -\frac{R}{r\sqrt{\alpha}}{}_c D_t^{1/2} v_i(t)$$

This formation with the leading coefficients specialized to one is the basis of semi-differential computing element. Fig. 3.7 gives a practical circuit for semi-integration with operational amplifiers.

3.5 Driving Point Impedance of Semi-infinite Lossy Transmission Line

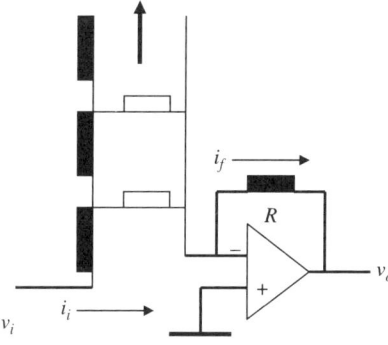

Fig. 3.6 Semi-differentiator

The circuit in Fig. 3.7 is to realize the fractional order PID analog control system. In this circuit, the offset adjustment parts are not explicitly shown. The semi-integral control will have transfer function as

$$\frac{V_0(s)}{V_i(s)} = \frac{Z_f}{Z_i} = \frac{\sqrt{R\frac{1}{Cs}}}{R} = \frac{K}{\sqrt{s}}.$$

By replacing s with $j\omega$, one gets the relation as

$$\frac{V_0(j\omega)}{V_i(j\omega)} = \frac{K}{\sqrt{\omega}} e^{-j\pi/4}.$$

This circuit behaves as constant phase element of angle $-45°$, meaning to an sinusoidal input the circuit will give a constant phase lag to the output by $45°$. By using values of the impedances, the transfer function constant

$$K = \frac{\sqrt{\frac{22 \times 10^3}{0.47 \times 10^{-6}}}}{22 \times 10^3} = 9.8,$$

and the transfer function is

$$\frac{V_0(j\omega)}{V_i(j\omega)} = 9.8\omega^{-0.5} e^{-j\pi/4}.$$

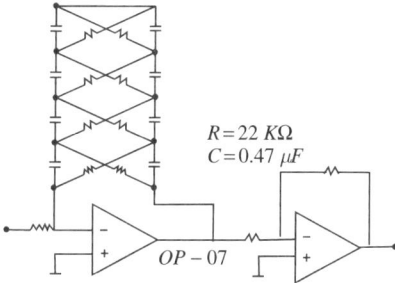

Fig. 3.7 Practical circuit for semi-integrator

Table 3.1 Practical results from semi-integrator circuit measurement

Input frequency (Hz) f	Input frequency (radian) ω	Phase angle (degree)	V_i (Volt)	V_0 (Volt)	$G = \frac{V_0}{V_i}$	$K = G\sqrt{\omega}$	$20\log(G)$ (dB)
50	314	−50.4	3.8	2.0	0.5263	9.32	−5.57
100	628	−45.0	3.8	1.5	0.3947	9.89	−8.07
150	942	−56.25	3.8	1.2	0.3158	9.69	−10.01
200	1257	−55.40	3.8	1.00	0.2632	9.33	−11.59
250	1571	−60.00	3.8	0.80	0.2105	8.19	−13.53
400	2513	−41.50	3.8	0.75	0.1974	9.89	−14.09
450	2827	−49.10	3.8	0.70	0.1842	9.79	−14.69
600	3770	−51.40	3.8	0.65	0.1710	10.49	−15.34
700	4398	−52.94	3.8	0.60	0.1578	10.46	−16.03
750	4712	−56.25	3.8	0.58	0.1526	10.47	−16.32
900	5655	−69.23	3.8	0.56	0.1473	11.07	−16.63
950	5969	−60.00	3.8	0.54	0.1421	10.9	−16.95

The practical results given in Table 3.1, where almost a constant phase is demonstrated (around $-55°$). The circuit is excited by sinusoidal voltage and the phase lag was recorded along with the peak–peak amplitude

3.5.2 Application of Fractional Integral and Fractional Differentiator Circuit in Control System

Analog or digital realization can give a control system design for fractional order control system. Figure 3.8 gives the block diagram representation of a classical integer order system (DC Motor) being controlled by a fractional order feedback controller.

System transfer function of DC motor is

$$G(s) = \frac{K}{Js(Ts+1)},$$

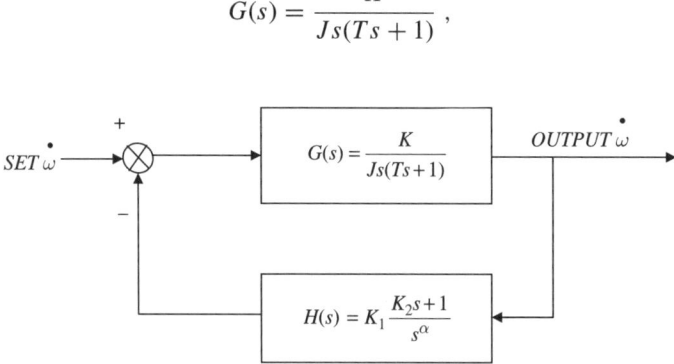

Fig. 3.8 Block diagram of fractional order control system

3.5 Driving Point Impedance of Semi-infinite Lossy Transmission Line

J is the payload (inertia). Phase margin of the controlled system is

$$\Phi_m = \arg |G(j\omega)H(j\omega)| + \pi,$$

the controller characteristics is

$$H(s) = K_1 \frac{K_2 s + 1}{s^\alpha}.$$

Here, choose $K_2 = T$. Note that $H(s)$ is composed of a differentiator of fractional order $(1 - \alpha)$ and an integral controller of order α. This gives constant phase margin as

$$\Phi_m = \arg |G(j\omega)H(j\omega)| + \pi$$
$$= \arg \left[\frac{K_1 K/J}{(j\omega)^{1+\alpha}} \right] + \pi = \arg \left[(j\omega)^{-(1+\alpha)} \right] + \pi = -(1+\alpha)\frac{\pi}{2} + \pi = \frac{1}{2}(\pi - \pi\alpha)$$

The close-loop transfer function is

$$G_c = \frac{G(j\omega)H(j\omega)}{1 + G(j\omega)H(j\omega)} = \frac{KK_1/J}{(s^{1+\alpha} + KK_1/J)}.$$

The step input response will be

$$y(t) = L^{-1} \left[\frac{KK_1/J}{s\left(s^{1+\alpha} + KK_1/J\right)} \right] = \left(\frac{KK_1}{J}\right) t^{1+\alpha} E_{1+\alpha, 2+\alpha} \left(-\frac{KK_1}{J} t^{1+\alpha}\right)$$

This is "iso-damping," meaning that the overshoot is same for various payloads (inertia); to have this type of control system is robust and efficient. Figure 3.9 gives the concept of iso-damping.

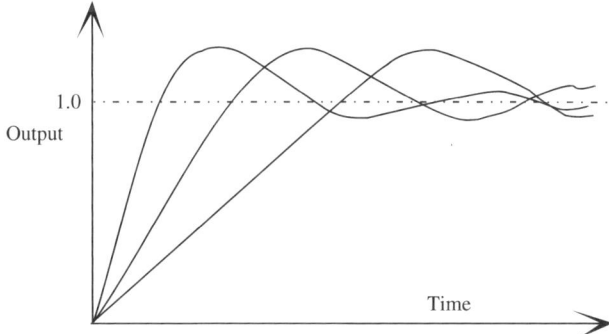

Fig. 3.9 Iso-damping in fractional order controlled system

3.6 Semi-infinite Lossless Transmission Line

Above discussion elaborates on semi-differentiation and semi- integration obtained for driving point impedance of semi-infinite lossy line. A lossless transmission line constitutes of L and C distributed throughout its length. In Fig. 3.4, the element L will replace R. The line considered here is the semi-infinite lossless line whose impedance is constant or an operator of zero order. The problem is written as

$$\frac{\partial^2 v(x,t)}{\partial t^2} = \frac{1}{LC}\frac{\partial^2 v(x,t)}{\partial x^2}, \ v(0,t) = v_1(t), \ v(\infty, t) = 0, \ v(x, 0) \& v'(x, 0)$$

given with

$$\frac{\partial i(x,t)}{\partial x} = -C\frac{\partial v(x,t)}{\partial t} \ \& \ \frac{\partial v(x,t)}{\partial x} = -L\frac{\partial i(x,t)}{\partial t},$$

where v is the voltage, i is the current, $v_I(t)$ is time-dependent input variable, L is inductance per unit length, and C is capacitance per unit length. A classical solution to this problem is obtained through iterated Laplace transforms as done for semi-infinite lossy line in Sect. 3.5. The main results are given below:

$$V(0,s) = \frac{-V^*(0,s)}{\sqrt{LC}s} + \frac{1}{\sqrt{LC}}[V(p,0)]_{p=s\sqrt{LC}} + \frac{1}{\sqrt{LC}s}[V'(p,0)]_{p=s\sqrt{LC}}$$

$$V^*(0,s) = \frac{dV(0,s)}{dx} \ \& \ V' = \frac{dv(x,0)}{dt}$$

This contains transfer function of driving point (not impedance) as

$$\frac{V(0,s)}{V^*(0,s)} = -\frac{1}{\sqrt{LC}s}$$

and in time domain

$$v(0,t) = -\frac{1}{\sqrt{LC}}\int \frac{dv(0,t)}{dx}dt + \varphi_1(t)$$

$\varphi_1(t)$ is the time-dependent initial condition. The transfer function consists of two parts: the forced response due to $V^*(0, s)$ and the initial condition response due to the initial voltage distribution in the lossless line. Using current expression as given,

3.6 Semi-infinite Lossless Transmission Line

the driving point impedance is obtained as follows:

$$V(0,s) = \sqrt{\frac{L}{C}} I(0,s) - \sqrt{\frac{L}{C}} \frac{[I(0,0)]}{s} + \frac{1}{\sqrt{LC}} [V(p,0)]_{p=s\sqrt{LC}}$$
$$+ \frac{1}{\sqrt{LC}s} [V'(p,0)]_{p=s\sqrt{LC}}$$

Notice that the voltage is composed of two parts: the forced response due to $I(0,s)$ and the initial condition response due to the initial voltage distribution in the lossless line. Considering only the first term, it can be seen that the impedance looking into this line is thus

$$Z(s) = \frac{V(0,s)}{I(0,s)} = \sqrt{\frac{L}{C}}$$

which is simply a constant. Mathematically, the impedance expressed in time domain as

$$v(0,t) = \sqrt{\frac{L}{C}} i(0,t) + \varphi_2(t)$$

has a time-dependent initial condition response due to initial voltage and current distribution and can be obtained by Laplace inverse of the last three terms of equation showing $V(0,s)$, $I(0,s)$ relationship, i.e.,

$$V(0,s) = \sqrt{\frac{L}{C}} I(0,s) - \sqrt{\frac{L}{C}} \frac{[I(0,0)]}{s} + \frac{1}{\sqrt{LC}} [V(p,0)]_{p=s\sqrt{LC}}$$
$$+ \frac{1}{\sqrt{LC}s} [V'(p,0)]_{p=s\sqrt{LC}}$$

Thus it can be seen that a simple constant gain operator (zero- order operator) can also have time-varying initial condition terms. Figure 3.10 gives the diagram of a lossless semi-infinite transmission line (a zero-order element). Though the order of operation is zero, i.e., it returns the input function (variable) unaltered (except gain or attenuation), yet in the theory of generalized calculus, the initial distributed charges and voltage stored will be returned to the output. This initial function is time varying into future. The initial conditions on the distributed L and C along the infinite line gives rise to initialization functions (of time). Note that this particular element (of zero order) does not call for differintegrations, but the initial conditions φ associated with this distributed characteristics is very important to generalized theory of initialized (fractional) calculus. Operational amplifier circuit realized with zero-order distributed element will give practical understanding for generalized (initialized) calculus; it is dealt in detail in Chaps. 6 and 7.

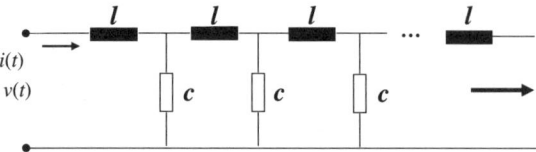

Fig. 3.10 Semi-infinite lossless transmission line

In Fig. 3.6, the input element is a lumped resistor R and the feedback element is a lumped capacitor C. Then this circuit configuration gives lumped integrator circuit. Let the input voltage $v_i(t)$ to the circuit be switched on at some time a, before the time $t = a$ the voltage is zero, and we start the circuit process (of integration) at time $t = c > a$. This implies that the capacitor is pre-charged with $q(c)$ Coulombs, from time a to c with initial voltage $v_o(c)$. This constant is thus the initial condition for this lumped element integrator circuit.

The describing equations for this configuration is as follows:

$$v_i(t) - 0 = i_f R$$

$$0 - v_o(t) = \frac{1}{C}\int_{t=c}^{t} i_f(t)dt + \frac{1}{C}\int_{t=a}^{t=c} i_f(t)dt = \frac{1}{C}\int_{c}^{t} i_f(t)dt + \frac{q(c)}{C}$$

$$= \frac{1}{C}\int_{c}^{t} i_f(t)dt + [0 - v_o(c)] = \frac{1}{C}{_c}D_t^{-1}i_f(t).$$

${_c}D_t^{-1}$ is integer order one integration process starting from time $t = c$ which includes the initialization, process that is charging of the capacitor C from time $t = a$ to $t = c$ which is represented as

$$\psi'(t) = \int_a^c i_f dt = \frac{1}{C}q(c) = -v_o(c) .$$

Therefore, the total process is un-initialized integration starting from time $t = c$, that is:

$$_c d_t^{-1} i_f = \int_c^t i_f dt$$

plus initialization integration process from a to c, that is:

$$_a d_c^{-1} i_f = \int_a^c i_f dt = \psi'(t)$$

3.6 Semi-infinite Lossless Transmission Line

These equations yield the final result by putting $i_i(t) = i_f(t)$

$$v_o(t) = -\frac{1}{RC}\int_c^t v_i(t)dt + v_o(c) = -\frac{1}{RC}{}_cD_t^{-1}v_i(t),$$

with $\psi(t) = -RCv_o(t)$. This is classical integer order calculus, with initialization as constant.

In the circuit of Fig. 3.6, we now replace the input element with semi-infinite lossless (LC) transmission line, a zero-order element, and the feedback element with lumped capacitor C_f. The transmission line terminal equation is re-written as

$$i(t) = \sqrt{\frac{C}{L}}v(t) + \varphi(t),$$

with $\varphi(t)$ as initial charge distribution on the distributed element.

The defining equations of this circuit are

$$i_i(t) = \sqrt{\frac{C}{L}}[v_i(t) - 0] + \varphi(t)$$

$$0 - v_o(t) = \frac{1}{C_f}\int_{t=c}^t i_f(t)dt + [-v_o(c)],$$

as done for the lumped integrator case above

$$i_i(t) = i_f(t)$$

Therefore solving for $v_o(t)$, we obtain

$$v_o(t) = -\left[\frac{1}{C_f}\sqrt{\frac{C}{L}}\right]\int_{t=c}^t v_i(t)dt - \frac{1}{C_f}\int_{t=c}^t \varphi(t)dt + v_o(c) = -\left[\frac{1}{C_f}\sqrt{\frac{C}{L}}\right]{}_cD_t^{-1}v_i(t)$$

where

$$\psi(t) = \sqrt{\frac{L}{C}}\int_c^t \varphi(t)dt + C_f\sqrt{\frac{L}{C}}v_o(c)$$

Here the initialization function is not a constant, but a function of time.

This expression is similar to the classical integer order integrator with lumped parameters as obtained earlier. Here the integrator is realized with distributed element. The important difference is the values of initialization function. For a distributed element integrator, the effect of past history is not only contained in a constant $v_o(c)$, which is charge on the capacitor C_f, but also carried in the remainder of the $\psi(t)$ function, which accounts for the distributed charge along the semi-infinite line. It is also observed here that the zero-order input element, as it is a wave equation

$$\frac{\partial^2 v(x,t)}{\partial t^2} = \frac{1}{LC} \frac{\partial^2 v(x,t)}{\partial x^2},$$

will simply propagate any perturbations in $v_i(t)$ along the semi-infinite line, never returning, thus never seen again, the only effect being proportional variations in the $i_i(t)$. This behavior is true for terminal charging. However for side charging (arbitrary charging with voltages on the distributed line), an additional time function may return to the circuit output, which is dependent on initial voltage distribution on the line.

The circuit of Fig. 3.6 when configured with input element as lumped capacitor C and the feedback element as lumped resistance R behaves as integer order differentiator. The constituent equations are

$$v_i(t) - 0 = \frac{1}{C} \int_{t=c}^{t} i_i(t)dt + \frac{1}{C} \int_{t=a}^{t=c} i_i(t)dt = \frac{1}{C} \int_c^t i_i(t)dt + \frac{q(c)}{C}$$

$$= \frac{1}{C} \int_c^t i_i(t)dt + v_i(c) = \frac{1}{C} {}_cD_t^{-1} i_i(t)$$

$$0 - v_o(t) = i_f R$$

$$i_i(t) = i_f(t)$$

This gives

$$v_o(t) = -RC \left[\frac{d}{dt} (v_i(t) - v_i(c)) \right] = -RC_c D_t^1 v_i(t) = -RC \left[{}_cd_t^1 v_i(t) + \psi(t) \right]$$

The initialization term

$$\psi(t) = \frac{d}{dt} v_i(c)$$

is taken normally as zero. However, the presence of initial charge in the input capacitor gives an impulse output at the start of differentiation process at $t = c$.

Modifying the circuit of Fig. 3.6 with input element as capacitor C_i and the feedback element with distributed LC zero-order element gives the integer order differentiator transfer character, with the concept of initialization function and generalized calculus. The defining equations are

$$v_i(t) - 0 = \frac{1}{C_i} \int_c^t i_i(t)dt + v_i(c) = \frac{1}{C_i} {}_cD_t^{-1} i_i(t)$$

3.6 Semi-infinite Lossless Transmission Line

For the distributed feedback zero-order elements, the expression in the circuit is

$$0 - v_o(t) = \sqrt{\frac{L}{C}} i_f(t) + \varphi(t) = \sqrt{\frac{L}{C}} i_f(t) + \sqrt{\frac{L}{C}} \psi(t)$$

Putting $i_i(t) = i_f(t)$, yields the final result as

$$v_o(t) = -\sqrt{\frac{L}{C}} \left[C_i \frac{d}{dt}(v_i(t) - v_i(c)) + \psi(t) \right] = -C_i \sqrt{\frac{L}{C}} \left[\frac{d}{dt} v_i(t) + \frac{1}{C_i} \psi(t) \right]$$

$$= -C_i \sqrt{\frac{L}{C}} {}_c D_t^1 v_i(t)$$

The generalized differentiation requires an initialization function. However for terminal charging case for integer order differentiation this initialization is zero but for side charged transmission line, an additional time function will be returned to the circuit output.

A simple gain (memory) less zero-order operator is realized by configuring the circuit of Fig. 3.6 with R_i as lumped resistor at input leg and R_f as lumped resistor at the feedback. The transfer characteristics will then be

$$v_o(t) = -\frac{R_f}{R_i} v_i(t) = -\frac{R_f}{R_i} {}_c D_t^0 v_i(t) = -\frac{R_f}{R_i} \left[{}_c d_t^0 v_i(t) + \psi(t) \right], \text{with} \psi(t) = 0,$$

clearly this circuit has no memory.

Zero-order circuit may be realized by employing semi-infinite distributed lossless transmission lines at input leg and one lumped resistor R at feedback, of circuit of Fig. 3.6.

The input leg equation with LC line is

$$i_i(t) = \sqrt{\frac{C}{L}} [v_i(t) - 0] + \varphi_i(t) = \sqrt{\frac{C}{L}} {}_c D_t^0 [v_i(t) - 0]$$

The feedback leg equation is

$$0 - v_o(t) = R i_f(t)$$

and $i_i(t) = i_f(t)$ gives

$$v_o(t) = -R i_f(t) = -R \sqrt{\frac{C}{L}} {}_c D_t^0 v_i(t) = -R \sqrt{\frac{C}{L}} \{ {}_c d_t^0 v_i(t) + \psi_i(t) \}$$

$$\text{where } \psi_i(t) = \sqrt{\frac{L}{C}} \varphi_i(t)$$

This zero-order operation in general returns the input function $v_i(t)$ (with amplification or attenuation), also provides the extra time function (associated with the memorized charges on the distributed element). This zero-order circuit has memory.

3.7 The Concept of System Order and Initialization Function

As the concept of order is central to the understanding of fractional (or integer) order systems, some discussion of this concept now follows. In this discussion, single input–output systems are considered. The examples in this chapter for heat flow and transmission line (lossy and lossless) gave the stage for half-order system or zero-order system. Recalling the characteristic equations or transfer function, we call a system first order, second order, third order, etc., similarly the system can be of fractional order too. We also consider that system representation is minimal and they are linear.

Mathematical order is defined as the highest derivative occurring in a given differential equation. The concept of mathematical order is applicable to both ordinary and fractional differential equations. Normally, when the order is used without qualifier, it implies the meaning of mathematical order.

For linear dynamic systems that are described by ordinary differential equations, the system mathematical order implies or is equivalent to the following:

1. The highest derivative in ordinary differential equation;
2. The highest power of Laplace variable s, in the characteristic equation;
3. The number of initializing constants required for the differential equation;
4. The length of the state vector;
5. The number of singularities in the characteristic equation;
6. The number of energy storage elements;
7. The number of independent spatial directions in which a trajectory can move;
8. The number of devices that can add $90°$ sinusoidal steady state phase lag;
9. The number of devises that retain some memory of the past.

The utility of the definition of mathematical order is that it infers all the system characteristics for the system, with integer order components.

Thus the benefit of having a definition for order for linear ordinary differential equations is that it allows a direct understanding of the behavior of given dynamic system. Unfortunately, for fractional differential equations, the order of the highest derivative does not infer all of the previously mentioned properties. Indeed, the most important characteristics of order in integer order ordinary differential equation is probably item (3), i.e., it indicates the number of initializing constants, which together with the differential equations allow prediction of the future behavior. In system terminology, this information provides initial states of the system being analyzed. For fractional order differential equations these indicators which are described to define 'order' of integer order differential equations, are not valid. The 'order' of fractional differential equation neither correspond to number of energy/memory storing elements, nor number of 'initializing constants' nor number of integrations (even fractional) required to solve the system. Thus the issue of order and the information required together with the fractional differential equation to predict the future is fundamental and should be treated differently.

In the examples of this chapter, it has been demonstrated that when specific differintegral operators ($q = 1/2$ and 0) are considered as semi-infinite systems, a time-dependent term resulting from the initial spatial condition should be added to the forced response. This is an important observation for solution of fractional differential equations that can have rather arbitrary initial conditions.

3.8 Concluding Comments

The practical examples in this chapter demonstrated the reality of the existence of fractional order differentiation and integrations, in natural description of systems. Interesting observations are obtained from analysis of semi-infinite systems; heat flow and current flow in lossy lines indicate the existence of semi-differintegration operations needed to describe transfer characteristics. Also realization of the transfer characteristics is possible by circuit synthesis and to have control system with robustness measure, independent of the gain. The "ifs and buts" regarding the definition of the order of the system for fractional differential equation is an open issue and cannot be directly related to integer order theory definitions. For example, a first-order system having say highest order of differentiation as unity shows under damped response to the step input excitation. Now the system looks as first-order system but with fractional order differentiation too will behave as though having some resonance. This behavior speaks that though the system may look classically first order yet due to fractional order terms present the behavior changes, so does the definition of the order. In this chapter, the examples point to the fact that distributed parameters do point toward fractional order system description, and in reality, the parameters are indeed distributed.

Chapter 4
Concept of Fractional Divergence and Fractional Curl

4.1 Introduction

Fractional kinetic equations of the diffusion are useful approach for the description of transport dynamics in complex systems, which are governed by anomalous diffusion and non-exponential relaxation patterns. The anomalous diffusion can be modeled by fractional differential equation, in time as well as space. For the spatial part, use of fractional divergence modifies the anomalous diffusion expression, in the modified Fick's law. Application of this fractional divergence is brought out in the nuclear reactor neutron flux definition. When anomalous diffusion is observed in time scale, the modification suggests the use of fractional kinetic equations.

Fractional curl operators will play perhaps role in electromagnetic theory and Maxwell equations. Here, example of electromagnetics is taken to have a feel how the fractional curl operator can map E and H fields in between the dual solutions of Maxwell equation.

4.2 Concept Of Fractional Divergence for Particle Flux

Because of relative simplicity and widespread use, the basis of local theory is discussed first. The local theory makes use of ADE (advection diffusion equation) as

$$\frac{\partial C}{\partial t} = \nabla.(-vC + D\nabla C) \tag{4.1}$$

where C is the solute concentration, and D and v are local dispersion and velocity tensors, respectively. The ADE is based on the classical definition of divergence of a vector field. The divergence is defined as the ratio of total flux through a closed surface to the volume enclosed by the surface when the volume shrinks toward zero.

$$div J \equiv \lim_{V \to 0} \frac{1}{V} \int_S J.ndS \tag{4.2}$$

where J is flux vector, V is an arbitrary volume enclosed by surface S, and n is unit vector normal to the surface.

This is valid only if the flux is indeed a "point" vector quantity relative to the scale of observation (e.g., heat flow in homogeneous material). Then the limit exists and the operator reduces the familiar dot product with gradient vector $[\partial/\partial x, \partial/\partial y, \partial/\partial z]$. Solute dispersion is counter example since it is primarily due to the velocity fluctuations that arise only as an observation space grows larger, invalidating the limit. The solute flux is due to combined effects of mean velocity (advection) and velocity fluctuation (dispersion). The dispersive fluxes for a given volume are typically averaged in some fashion (volumetric, statistical) and approximated by Fick's first law. Since velocity itself is a variable function of space, as control volume shrinks (as divergence requires), the velocity fluctuations and the dispersive flux disappear. Therefore true divergence of the macroscopic solute flux cannot contain a macroscopic dispersive term.

Because of the limit in (4.2), the classical Gauss divergence theorem discounts macro-dispersion until a point vector can approximate the dispersive flux. This calls for separation of scales: The scale of the transport process must be much larger than some finite volume at which the ratio in (4.2) becomes seemingly constant. For these things to happen, the dispersive flux must not increase as volume passes some largest size. This representative elementary volume (REV) for dispersion is the point at which the deviations in the velocity field are negligible. The divergence is associated with a non-zero volume and is given by the first derivative of total surface flux to volume (Fig. 4.2) rather than the limit of the derivative at zero volume. The dispersion coefficient does not grow (scale), if the ratio of the surface flux to volume is constant over some range of volume (Fig. 4.1 and *stepped solid lines* of Fig. 4.2b make piecewise constant slope of Fig. 4.2a). Therefore at some larger scale of observation (non-zero), the ratio of the total surface flux to volume is constant over large range of arbitrary volumes, and the relatively constant first derivative (the de facto divergence) allows assignment of dispersion coefficient. Both volume averaging and ensemble averaging concepts are based on this idea of separation (or distinction) of scale.

At the field scale, at least two problems occur that make it difficult to rely on the REV method. First, even if there is a distinct hierarchy, the act of measurement involves a volume integration, which impacts the dispersion coefficient. Second,

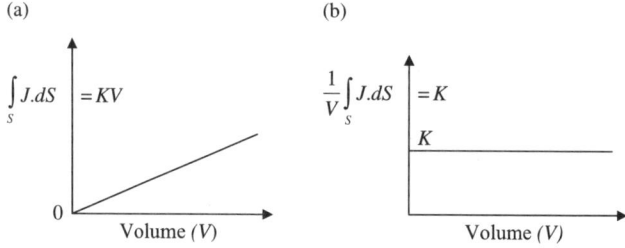

Fig. 4.1 Classical definition of divergence of flux vector

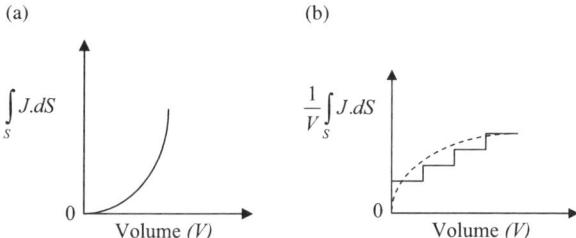

Fig. 4.2 Effect of dipersive flux and neutron velocity fluctuation with macroscopic scale of observation extension

there is a long-standing and growing body of evidence that real solute materials have evolving heterogeneity. If that is the case, then there will be no separation of scales, and the *dashed line* of Fig. 4.2b represents the flux, as continuous one instead of stepped one.

An integer order divergence theorem forces a scaling parameter, since ratio of flux to volume is scale dependent. Rather than to use a step function approximation (Fig. 4.2b), of the growth of dispersive flux with scale, which forces D to take on increasing values, one might try to describe the evolving *dashed line* (Fig. 4.2b). Non-local including convolutional theories do this by integrating cumulative effects of dispersion over any length scale and/or time scale. A subset of these uses the mathematical tools of fractional calculus, which are non-local operators for fractal functions.

4.3 Fractional Kinetic Equation

The fundamental laws of physics are written as equations for the time evolution of a quantity $X(t)$ with $dX(t)/dt = -AX$, where this could be Maxwell's equation, Schrodinger's equation, Newton's law of motion, or Geodesic equations. The mathematical solution for the linear operator A is $X(t) = X(0) \exp\{-At\}$, putting $\tau = A^{-1}$ the standard exponential relaxation law is

$$X(t) = X(0) \exp\{-t/\tau\}. \tag{4.3}$$

Complex systems and investigations of their structural and dynamical properties have been established on the physics agenda. These structures with variations are characterized through

a. a large diversity of elementary units;
b. strong interactions between the units; and
c. a non-predictable or anomalous evolution in course of time.

Complex systems and their study play a dominant role in exact and life sciences, embracing a richness of systems such as glasses, liquid crystals, polymers, proteins, biopolymers, or even ecosystems. In general, the temporal evolution of, and within, such systems deviates from the corresponding standard laws. With the development of higher resolution experiments, these deviations have become more prominent.

One can have stretched exponential behavior as

$$X(t) = X(0) \exp\{-(t/\tau)^\alpha\} \tag{4.4}$$

with $0 < \alpha < 1$, or one can visualize the asymptotic power law as

$$X(t) = X(0)(1 + t/\tau)^{-n}, \text{ with } n > 0 \tag{4.5}$$

Similarly the diffusion process in various complex systems usually no longer follow Gaussian statistics, and thus Fick's second law fails to describe the related transport behavior. Especially one observes deviations from the linear dependence of mean squared displacement:

$$< x^2(t) > \approx K_1 t \tag{4.6}$$

which is characteristic of Brownian motion, and such a direct consequence of central limit theorem and the Markovian nature of the underlying stochastic process. Instead anomalous diffusion is found in a wide diversity of systems, its hallmark being non-linear growth of the mean squared displacement in course of time, following power law:

$$< x^2(t) > \approx K_\alpha t^\alpha \tag{4.7}$$

which is ubiquitous to a diverse number of systems. There exists a variety of other patterns, such as logarithmic time dependence. The anomalous diffusion behavior manifested in (4.7) is intimately connected with break down of the central limit theorem, caused by either broad long-tailed distributions or long-range correlations. These broad spatial jumps or waiting times distributions lead to non-Gaussian and possibly non-Markovian time evolution way of diffusion, manifesting into non-local temporal phenomena. Note that the unit of diffusion coefficient in (4.7) is having unit $[K_\alpha] \equiv cm^2 s^{-\alpha}$, according to anomalous diffusion exponent α. This exponent if $0 < \alpha < 1$ defines sub-diffusive transport, and it defines super-diffusive phenomena for $1 < \alpha < 2$. For $\alpha = 1$, the transport phenomena is normal integer order and Fickian.

Standard integer order kinetic equation when integrated gives the following:

$$X_i(t) - X_0 = -c \int_0^t X(t) = -c_0 D_t^{-1} X(t) \tag{4.8}$$

$_0D_t^{-1}$ is the standard Riemann–Liouville integral operator. The number density of the species i, $X_i = X_i(t)$ is a function of time, and $X_i(t=0) = X_0$ is the number density of species i at time $t=0$. If we drop the index i in (4.8) and replace c by c^α, then the solution of the generalized fractional order diffusion equation:

$$X(t) - X_0 = -c^\alpha {}_0D_t^{-\alpha} X(t) \tag{4.9}$$

is

$$X(t) = X_0 \sum_{k=0}^{\infty} \frac{(-1)^k (ct)^{k\alpha}}{\Gamma(k\alpha + 1)} \tag{4.10}$$

It can be written as compact form, by use of Mittag-Leffler function as

$$X(t) = X_0 E_\alpha(-c^\alpha t^\alpha) \tag{4.11}$$

4.4 Nuclear Reactor Neutron Flux Description

The neutron balance description in nuclear reactor is defined by transport theory. The basic transport equations are then approximated by several coupled differential equations. One of the simplified approximation of the reactor representation given to engineers is the neutron diffusion equation sets in multi-energy group or single-energy group. In all these diffusion equations, the leakage term has Fick's law of diffusion, where the neutron flux is assumed to be a point quantity. For larger reactor representation, several of these diffusion equations are formed and modeled by region to region coupling coefficients. Engineering science then proceeds on these approximates to obtain reactor transfer function model, and then various control system analyses are done. For complex systems, the integer models of the reactor may not suffice and thus a fractional order model for obtaining flux profile or kinetics may describe the complex reality better. The argument is similar to that described for heat transfer model (Chap. 2.4), where distributed and complex parametric spreads and size factor are described better by fractional transient heat transfer equation.

4.5 Classical Constitutive Neutron Diffusion Equation

In classical sense, the constitutive equation assumes point neutron flux, with v as the average speed of the neutrons passing through an area with n neutrons per unit volume as neutron density. The vector quantity representing the neutron flux is J. The following will elucidate the classical statements.

$$\vec{J} = n\mathbf{v}$$
$$\phi = nv$$
$$\vec{J} = -D\nabla\phi$$

Consider a closed volume, the loss of neutrons from the closed surface is given as surface integral of neutron current, $J.dS$. The loss occurring in the volume by absorption is given by absorption cross-section and then taking volume integral of $\Sigma_a \phi dV$. This total loss, when equated to the source term, gives the classical constitutive neutron diffusion equation, as depicted below.

$$\int_S J.dS + \int_V \Sigma_a \phi dV = \int_V S dV$$

The above integral form when converted to volume integral is:

$$\int_V (\nabla.J + \Sigma_a \phi - S) dV = 0 \quad \text{or} \quad \nabla.J + \Sigma_a \phi - S = 0 \text{ for equilibrium}$$

Using the expression of $J = -D\nabla\phi$, we obtain the following:

$$-D\nabla^2 \phi + \Sigma_a \phi - S = 0$$

In the steady state, the RHS of above constitutive equation is zero, and if there is time changing flux, then that is put in the RHS as

$$D\nabla^2 \phi - \Sigma_a \phi + S = \frac{\partial n}{\partial t} = \frac{1}{v}\frac{d\phi}{dt}$$

4.5.1 Discussion on Classical Constitutive Equations

The classical neutron diffusion constitutive equation as described is based on the classical divergence of the divergence of a vector field. The divergence is defined as the ratio of total flux through a closed surface to the volume enclosed by the surface, when volume shrinks toward zero.

$$div J = \lim_{V \to 0} \frac{1}{V} \int_S J.dS$$

Where J is the flux vector, V is an arbitrary volume enclosed by surface S. The dot product of vector J with the surface dS is obvious; this is valid only if the flux is indeed a "point" quantity relative to the scale of observation. Neutron diffusion is counterexample; this is primarily, due to velocity fluctuations (even at constant

4.5 Classical Constitutive Neutron Diffusion Equation

energy/temperature) that arise only as the observation space grows larger, invalidating the limit. Also the neutrons are no longer in homogeneous medium. The dispersive fluxes for a given volume are typically averaged in some fashion (volumetric, statistical) and are approximated by Fick's first law as we have obtained in deriving the classical constitutive equation for neutron diffusion, $\vec{J} = -D\nabla\phi$. As the control volume shrinks to zero, the velocity fluctuations and the dispersive flux disappear. Therefore in a true sense, the classical divergence theorem discounts the real effects of macroscopic in-homogeneity and the fluctuations associated with neutron diffusion in a reactor.

Because of the limit in the divergence definition, the classical Gauss divergence theorem discounts the effect of large volume until the dispersive flux can be approximated by a point quantity.

4.5.2 Graphical Explanation

Refer Fig. 4.1. In Fig. 4.1a, it is shown that the surface flux with respect to volume of the observation space is of constant slope line. Figure 4.1b plots the ratio of the surface flux with respect to the control volume (first derivative of Fig. 4.1a).

Figure 4.1 shows, in simplistic manner, that if the surface flux of neutrons with average constant velocity grows in linear fashion with respect to the volume of the observation space, then in this case, the ratio of the surface flux to the control volume remains fixed. In this particular (ideal) case, making the control volume shrink to zero will yield ideal definition of the divergence of the vector flux (neutron current density). This simplistic picture neglects the effect of in-homogeneous medium and macroscopic dispersion, fluctuating velocity effects and effects due to neighborhood, neutron currents.

Figure 4.2 is an extension of Fig. 4.1 showing the macroscopic effects of surface flux manifestaion as the control volume is enlarged. The observation space when enlarged as shown in Fig. 4.2b captures dispersive effect of neutrons as magnified by the staircase type of ratio of surface flux to the volume figure. The effect can be seen as surface flux gets manifested as some power of observation space (volume). Figure 4.2b is the first derivative of Fig. 4.2a and shows that at quasi large observation space (control volume), one gets seemingly constant ratio of surface flux to volume, therefore yielding a non-local divergence of the neutron flux vector. This definition of non-local divergence is what contradicts the classical divergence, where the control volume is made to shrink to zero.

4.5.3 About Surface Flux Curvature

Refer Fig. 4.2a. The curvature is concave in nature as the observation space (control volume) is made bigger. Contention could have been that why the curvature is taken as concave instead of convex. Here some practical reasoning will elucidate the nature of the curve shown in Fig. 4.2. For a very small observation space area, the surface flux is the product of neutron current and that area. As the area is made larger, the neighbouring neutrons effect the neutron current in the wider area of measurement. This gives the larger value of the neutron current for the newer area considered. This increment in the neutron current is what gets integrated in the surface integral giving the concave shape (Fig. 4.2a).

This is elaborated in Fig. 4.3. Let the observation surface area for measurement of neutron surface flux be divided into squares as shown. Assume that each center of the square is having one neutron. If all the neutrons are at rest without any velocity fluctutaions, then there will not be any finite probabilty that it may jump across to the next adjacent box. However, the case is not so, as there always is a finite probabilty of having neutrons designated for a particular box finding into the adjacent box. However, if the area of observation is very small as depicted by smaller circles inside each box, the fluctutaion effect of neutron velocity will not be observed. Therefore with the smaller circles in the observation space measures a smaller neutron current (solely due to the presence of its own neutron in the squrae box). However, the observation area is made into larger circles as shown in Fig. 4.3. Here we see that with enlarged area the effect of neutrons in the adjacent square will enhance the neutron current compared to the first smaller area. Also this bigger circle will catch the effect of velocity fluctuations and therefore will show larger magnitude of neutron current. This simplistic explanation is justified for the observation that the shape of the surface flux with respect to the observation space (Fig. 4.2) is concave and not convex.

4.5.4 Statistical and Geometrical Explanation for Non-local Divergence

Figure 4.3 divides the space into grids. The fluctuations in velocity cause the violation of classical limit of volume shrink to zero, for classical divergence also

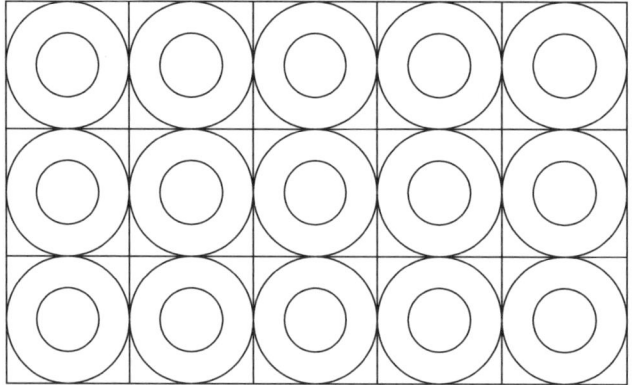

Fig. 4.3 The effect of growing observation space modifying the neutron current

elucidated by the fact that at a particular space the neutrons will have spatial long-tailed distributions. The effect of this long-tailed statistical probability distribution will thus get enhanced by the use of non-local divergence, and this reality effect will thus be shown with avoidance of volume shrinkage to zero. Also in reality, the coupling between various zones in the reactor takes place. The non-local divergence with the principle of non-zero volume therefore is the apt tool for constitutive equation for neutron diffusion equation for reactor description.

Refer the classical neutron diffusion equation and Fig. 4.2. The surface flux is $\int_S J.dS$; then the ratio of the surface flux to the volume is taken to consider the leakage through surface term as $\nabla.J$, at volume shrinking towards zero. Then with Fick's law, we get the leakage of neutrons through a closed surface as $\nabla.(-D\nabla\phi) = -D\nabla^2\phi$. This term in one dimension case is

$$-D\frac{d^2\phi}{dx^2}$$

Examining the above and relating this to Fig. 4.2, one may say that divergence is the slope of the surface flux (Fig. 4.2a). If this curvature has square law variation in the shape, then the double derivative will be constant, and we have the entire curvature captured in that constant. If the curvature of the surface flux (Fig. 4.2a) is not having $\approx x^2$ variation, but say has $\approx x^{1.5}$ variation, then double derivative will not capture the information about the curvature in a constant.

4.6 Fractional Divergence in Neutron Diffusion Equations

The neutron leakage through closed surface enclosed by finite control volume may be changed to $-D\frac{d^\beta\phi}{dx^\beta} = \frac{\partial^\alpha}{\partial x^\alpha}.(-D\frac{d\phi}{dx})$. The bracket quantity is Fick's first law. Here β is between 1 and 2, and α is between 0 and 1, non-integer fractional real numbers. In the divergence formulations, the vector

$$[\frac{\partial^\alpha}{\partial x^\alpha}, \frac{\partial^\alpha}{\partial y^\alpha}, \frac{\partial^\alpha}{\partial z^\alpha}] \to \nabla^\alpha,$$

the fractional divergence operator, which acts on the neutron current vector J.

The constitutive equation in fractional divergence form may thus be as follows:

$$\nabla^\alpha.J + \Sigma_a\phi - S = 0$$
$$\nabla^\alpha(-D\nabla\phi) + \Sigma_a\phi - S = 0$$
$$D\nabla^{1+\alpha}\phi - \Sigma_a\phi + S = 0$$

Converting to one-dimensional form to get neutron flux profile in a reactor or neutron flux profile near a source in scattering medium, we will look at solutions of the form as follows:

$$D\frac{d^{1+\alpha}\phi}{dx^{1+\alpha}} - \Sigma_a\phi + S = 0, \ldots 0 < \alpha < 1$$

$$D\frac{d^{\beta}\phi}{dx^{\beta}} - \Sigma_a\phi + S = 0, \ldots 1 < \beta < 1$$

One may interpret the simplified form of $\nabla^{\alpha}.J$ is that a fractional divergence operator is applied to Fickian dispersion term. For illustration of how fractional derivatives relate to the definition of divergence in neutron transport, consider two simple functions $f(x) = x^2$ and $f(x) = x^{1.5}$. Recalling the fundamental of the derivatives, we recall that the derivatives of function give the idea of the curvature contained in the curve. Successive derivatives will strip of the independent variable and subsequently will show up the curvature contents. Take the example $f(x) = x^2$, the first and second derivatives are $f^{(1)}(x) = 2x$ and $f^{(2)}(x) = 2$, respectively. In this example, the second derivative contains all the information about the function and that is constant (i.e., 2). This argument favors well for well-behaved integer order functions. Now let us change the function and take any real order function, say $f(x) = x^{1.5}$; the first and second derivative is $f^{(1)}(x) = 1.5x^{0.5}$ and $f^{(2)}(x) = 0.75x^{-0.5}$. In both the derivatives, for this real ordered function, the derivatives vary with the independent variable. Refer Fig. 4.2b; the effect of growing control volume is approximated as steps. Each step signifies growing value of D. The shape emerged as dotted curve in Fig. 4.2b, approximates the first derivative of Fig. 4.2a. Here the first derivative or any other higher order integer derivative fails to contain the curvature information.

In the functions with real order (other than integer order), if the concepts of fractional calculus are applied then we can contain the curvature information of $f(x)=x^{1.5}$ by taking 1.5 derivative of the $f(x)$, symbolically $d^{1.5}x^{1.5}/dx^{1.5}=1.33$.

$$\frac{d^{\alpha}}{dx^{\alpha}}x^u = \frac{\Gamma(u+1)}{\Gamma(u-\alpha+1)}x^{u-\alpha}$$

$$\frac{d^{1.5}}{dx^{1.5}}x^{1.5} = \frac{\Gamma(1.5+1)}{\Gamma(1.5-1.5+1)}x^{1.5-1.5} = \Gamma(2.5) = 1.33$$

Therefore if the fractional differential operator is chosen in which the fractional order of differentiation matches the power law scaling of the function, then the curvature is reduced to a constant and all the scaling information is contained in the order of the derivative and that constant. If the neutron plume is traveling through material with evolving heterogeneity, then a fractional divergence might account for the increased dispersive flux over larger range of measurement scale.

This argument sets the stage for writing neutron diffusion equation with fractional calculus. The above derivation and discussion completes the description of neutron diffusion equations in fractional calculus.

In deriving the fractional differential equations for neutron diffusion in an enclosed volume, we argued the basis of taking a larger observation space for defining divergence. The non-local formation of the divergence thus gives the effect

4.6 Fractional Divergence in Neutron Diffusion Equations

of macroscopic effects caused by velocity fluctuations, the coupling effect of nearby zonal neighborhood neutrons (refer Fig. 4.3). This therefore is making the constitutive descriptive equations closer to reality. The classical integer order constitutive differential equations are approximations to everything as "point" quantity in time or in space. The classical integer order methods do not thus take into account the space history or time history and therefore cannot represent the natural laws close to reality. Fractional calculus does take of all these reality and therefore is more appropriate for representation of natural phenomena. Refer Fig. 4.3, as an outside observer, let us try to visualize space squares as depicted in the figure. The squares without any neutrons in them look different to an outside observer as compared to the squares with the neutrons, while observer sitting in the same squares will not notice the difference in the squares with or without neutrons. So the observer in the same space will apply point quantity and will try to describe the neutron balance by classical integer order calculus. Whereas to the outside observer the squares or the space will appear transformed with or without the presence of neutrons. The outside observer thus will apply this space transformation correction factor and obtain some different results, and that result will be close to reality.

4.6.1 Solution of Classical Constitutive Neutron Diffusion Equation (Integer Order)

This section will serve as a revision to simple classical solution of the diffusion equation. Then in the next section, we will solve the fractional differential equation obtained. This we will demonstrate the space variables in one dimension for simplicity.

$$D\nabla^2 \phi - \Sigma_a \phi + S = \frac{dn}{dt} = \frac{1}{v}\frac{d\phi}{dt}$$

$$S = k_\infty \Sigma_a \phi$$

$$D\nabla^2 \phi + (k_\infty - 1)\Sigma_a \phi = \frac{1}{v}\frac{d\phi}{dt}$$

The flux term is variable of space and time. The source term multiplication law governs S. The separation of variables will give the following for the flux term which can be substituted in the basic constitutive equation, and following expressions will emerge.

$$\phi = \phi(r)e^{-\Lambda t}$$

$$D\nabla^2 \phi(r) + (k_\infty - 1)\Sigma_a \phi(r) = \frac{-\Lambda}{v}\phi(r)$$

Λ is positive for sub-critical, negative for super critical, and zero for critical equilibrium reactor. We replace the space coordinate r, by x and with substitution of B as geometric buckling, we get following simple form. The temporal solution is avoided for simplicity.

$$B^2 = \frac{(k_\infty - 1)\Sigma_a + \frac{A}{v}}{D}$$

$$\frac{d^2\phi(x)}{dx^2} + B^2\phi(x) = 0$$

Here we can apply standard Laplace method with initial conditions at $x = 0$ at the center point of the reactor geometry having constant flux and at the walls at $x = a$ zero flux. General Laplace formula for derivative of function is indicated below and is applied to have polynomial form.

$$s^2\Phi(s) - \sum_{k=0}^{1} s^k \frac{d^{2-k-1}\phi(x)}{dx^{2-k-1}}]_{at\ x=0} + B^2\Phi(s) = 0$$

$$s^2\Phi(s) - \frac{d\phi(x)}{dx}]_{at\ x=0} - s\phi(x)]_{at\ x=0} + B^2\Phi(s) = 0$$

$$\frac{d\phi(x)}{dx}]_{at\ x=0} = 0 \text{ and } \phi(x)]_{at\ x=0} = C$$

The above initial condition gives simple equation as $s^2\Phi(s) - sC + B^2\Phi(s) = 0$

$$\Phi(s) = \frac{sC}{s^2 + B^2}$$

taking the inverse $\phi(x) = C \cos Bx$

4.6.2 Solution of Fractional Divergence Based Neutron Diffusion Equation (Fractional Order)

With the extension of the above method, we try to solve the fractional differential equation:

$$\frac{d^\beta\phi(x)}{dx^\beta} + B^2\phi(x) = 0$$

$$1 < \beta < 2$$

$$L\left(\frac{d^\alpha f(x)}{dx^\alpha}\right) = s^\alpha F(s) - \sum_{k=0}^{n-1} s^k \frac{d^{\alpha-k-1} f(x)}{d^{\alpha-k-1}}]_{at\ x=0}$$

$$s^\beta \Phi(s) - \frac{d^{\beta-1}\phi(x)}{dx^{\beta-1}}]_{at\ x=0} - s\frac{d^{\beta-2}\phi(x)}{dx^{\beta-2}}]_{at\ x=0} + B^2\Phi(s) = 0$$

4.6 Fractional Divergence in Neutron Diffusion Equations

The above is Laplace transformation for LHD definition of the fractional derivative. In this expression, the second and the third term of the left hand side has fractional derivative of the flux at initial point, which is physically difficult to define and to realize the same by experimental measurements is difficult at this stage. Let us try to make use of Laplace transformation of RHD Caputo definition, as given below:

$$L\left(\frac{d^\alpha}{dx^\alpha} f(x)\right) = s^\alpha F(s) - \sum_{k=0}^{n-1} s^{\alpha-k-1} \frac{d^k}{dx^k} f(x)_{at\ x=0}$$

$$s^\beta \Phi(s) - s^{\beta-1} \phi(x)_{at\ x=0} - s^{\beta-2} \frac{d}{dx} \phi(x)_{at\ x=0} + B^2 \Phi(s) = 0$$

We relate the above expression physically to the earlier initial condition taking second term as C and third term as zero as done in Sect. 4.6.1. Here the integer order derivative comes as initial condition, therefore physically realizable from measurements and observations.

$$\Phi(s) = \frac{s^{\beta-1} C}{s^\beta + B^2}$$

$$\phi(x) = C.L^{-1}\left[\frac{s^{\beta-1}}{s^\beta + B^2}\right]$$

The solution of the fractional differential equation for the constitutive neutron balance equation therefore is with Laplace identity

$$L(E_\alpha(-\lambda t^\alpha)) = \frac{s^{\alpha-1}}{s^\alpha + \lambda},$$

we obtain

$$\phi(x) = C.E_\beta(-B^2 x^\beta) = C \sum_{k=0}^{\infty} \frac{(-B^2 x^\beta)^k}{\Gamma(\beta k + 1)}$$

$$\phi(x) = C + C\frac{(-B^2 x^\beta)}{\Gamma(\beta+1)} + C\frac{(B^4 x^{2\beta})}{\Gamma(2\beta+1)} + C\frac{(-B^6 x^{3\beta})}{\Gamma(3\beta+1)} + \cdots$$

The above is flux mapping obtained by the solution of fractional order neutron constitutive equations that is obtained by the concept of fractional divergence.

Let us see what classical flux pattern and fractional order flux pattern are same when we take the fractional order equal to 2, the integer order.

Solution in the classical form is cosine function and series representation of the same is

$$\cos(x) = 1 - \frac{x^2}{2!} + \frac{x^4}{4!} - \frac{x^6}{6!} + \ldots$$

$$\beta = 2, \phi_{\beta=2}(x) = CE_\beta(-B^2x^2) = C\sum_{k=0}^{\infty} \frac{(-B^2x^2)^k}{\Gamma(2k+1)}$$

$$= C\left[1 - \frac{(B^2x^2)}{\Gamma(3)} + \frac{(B^2x^2)^2}{\Gamma(5)} - \frac{(B^2x^2)^3}{\Gamma(7)} + \ldots\right]$$

$$\Gamma(n+1) = n!$$

$$\phi_{\beta=2}(x) = C\left[1 - \frac{(Bx)^2}{2!} + \frac{(Bx)^4}{4!} - \frac{(Bx)^6}{6!} + \ldots\right] \approx C.\cos(Bx)$$

Therefore when the fractional order equals the integer order, we get classical flux profile. This is proof of our assumption that indeed neutron flux being not a point quantity be represented as fractional divergence of order less than unity.

4.6.3 Fractional Geometrical Buckling and Non-point Reactor Kinetics

The above concept of fractional divergence gave a deviation from ideal flux map (cosine). The term geometrical buckling is indicative of the flux profile of neutron flux inside the reactor. Measuring the actual flux distribution and then controlling the power of reactor is one mode of reactor control. Now if the control computer is kept with a map of cosine table and the neutron spatial detectors are mapping in each control cycle, then a deviation, the unwarranted correction cycles, will keep the control devises moving. Actually the correction may not be called for if the control computer is programed with actual fractional geometrical buckling data. The fractional divergence has given the new thought of "fractional geometrical buckling," which in turn when used with basic multiplying factor k_∞ gives rise to a concept of fractional criticality. The describing reactor kinetics with fractional divergence will give the concept of non-point kinetic description.

4.7 Concept of Fractional Curl in Electromagnetics

Fractional curl operator has been utilized to find the new set of solutions to Maxwell's equations by fractionalizing the principle of duality. New set of solutions is named as fractional dual solutions.

4.7.1 Duality of Solutions

The electromagnetic theory is based on the principle of duality. Dual solutions of electromagnetics means any solution to a problem containing electric source can be converted into a dual solution to the problem containing a magnetic source. Duality does arise in circuit theory as circuit of Thevenin voltage source can be converted to Norton current source. In electromagnetics, the set $(E, H, D, B, \mu, \varepsilon)$ has dual solution with set $(H, -E, B, -D, \varepsilon, \mu)$. This can be demonstrated by example as replacing $E \to H$, $H \to -E$, $\mu \to \varepsilon$, and $\varepsilon \to \mu$ in $\nabla \times E = -j\omega H$ and $\nabla \times H = j\omega E$, keeps the two Maxwell equations same.

4.7.2 Fractional Curl Operator

Fractional curl operator has been utilized to find the new set of solutions to Maxwell's equations by fractionalizing the principle of duality. In electromagnetics, the principle of duality states that if $(E, \eta H)$ is one set solutions (original) to Maxwell equations, then other set of solutions (dual to original) is $(\eta H, -E)$, where η is the impedance of the medium. The solution that may be regarded as intermediate step between the original and dual to the original solutions may be obtained using the following relationship.

$$E_{fd} = \frac{1}{(jk)^\alpha}(\nabla\times)^\alpha E \text{ and } \eta H_{fd} = \frac{1}{(jk)^\alpha}(\nabla\times)^\alpha H,$$

where $(\nabla\times)^\alpha$ means the fractional curl operator and $k = 2\pi f \sqrt{\mu\varepsilon} = \omega\sqrt{\mu\varepsilon}$ is the wave number of the medium. The subscripted E_{fd}, H_{fd} notations mean fractional dual solutions. Only unbounded medium (with no reflection boundaries) will be considered in this book for demonstration of fractional curl operator. However, standing wave patterns in reflection media and other studies of transmission line wave propagation are possible.

4.7.3 Wave Propagation in Unbounded Chiral Medium

In a chiral medium, electric flux and magnetic flux densities are composite quantity and are represented as $D = \varepsilon[E + \beta\nabla \times E]$ and $B = \mu[H + \beta\nabla \times H]$. Therefore the D (and B) at any point x depends on the electric field at other point x. This spatially dispersive property is non-local property and here the fractional calculus is used. The factor β is the chirality property of the media.

Consider a uniform plane wave propagating in z-direction in an unbounded lossless isotropic chiral medium. According to field decomposition, field quantities E and H may be thought of as two plane waves i.e. (E_+, H_+) and (E_-, H_-). The electric field corresponding to two-wave field is $E_\pm(z) = E_\pm(0)\exp(jk_\pm z)$, where $k_\pm = k(1 + \kappa_r)$ is wave number of the two wave fields, $k = \omega\sqrt{\mu\varepsilon}$ and $\kappa_r = \kappa\sqrt{\mu_0\varepsilon_0/\mu\varepsilon}$. Using the following relation

$$\eta_{\pm} H_{\pm}(z) = \pm j E_{\pm}(z)$$

corresponding magnetic field may be obtained. In the above expression
$\eta_{\pm} = \sqrt{\frac{\mu_{\pm}}{\varepsilon_{\pm}}} = \eta$. This means each wave field sees media with equivalent constitutive parameters as $(\varepsilon_{+}, \mu_{+})$ and $(\varepsilon_{-}, \mu_{-})$. Medium parameters of the equivalent isotropic media are related to the parameters of this chiral medium by
$\varepsilon_{\pm} = \varepsilon(1 \pm \kappa_r)$ and $\mu_{\pm} = \mu(1 \pm \kappa_r)$. Simple expressions of wave fields can be written as

$$E_{+} = \frac{1}{2}(E - j\eta H)$$
$$E_{-} = \frac{1}{2}(E + j\eta H)$$
$$H_{+} = \frac{1}{2}\left(H + j\frac{E}{\eta}\right)$$
$$H_{-} = \frac{1}{2}\left(H - j\frac{E}{\eta}\right)$$

The total field in chiral medium is

$$E(z) = E_{+}(0) \exp(jk_{+}z) + E_{-}(0) \exp(jk_{-}z)$$
$$\eta H(z) = j[E_{+}(0) \exp(jk_{+}z) - E_{-}(0) \exp(jk_{-}z)].$$

Fractionalizing the electric field
$E_{+}(z) \& E_{-}(z)$, we get

$$\begin{aligned} E_{fd+}(z) &= \frac{1}{(jk)^{\alpha}} (\nabla \times)^{\alpha} E_{+}(z) \\ &= \frac{1}{(jk)^{\alpha}} \{(z \times)^{\alpha} E_{+}(0)\} \left\{ \frac{d^{\alpha}}{dz^{\alpha}} \exp(jk_{+}z) \right\} \\ &= E_{+}(0) \exp\left\{ j\left(k_{+}z + \frac{\alpha\pi}{2}\right) \right\} \end{aligned}$$

Similarly, $E_{-}(z) = E_{-}(0) \exp\left\{ j\left(k_{-}z - \frac{\alpha\pi}{2}\right) \right\}$

Fractional dual fields corresponding to the original field $E(z)$, $H(z)$ may be written as

$$E_{fd}(z) = E_{+}(0) \exp\left\{ j\left(k_{+}z + \frac{\alpha\pi}{2}\right) \right\} + E_{-}(0) \exp\left\{ j\left(k_{-}z - \frac{\alpha\pi}{2}\right) \right\}$$
$$\eta H_{fd}(z) = j[E_{+}(0) \exp\left\{ j\left(k_{+}z + \frac{\alpha\pi}{2}\right) \right\} - E_{-}(0) \exp\left\{ j\left(k_{-}z - \frac{\alpha\pi}{2}\right) \right\}]$$

It is obvious that for $\alpha = 0$, $E_{fd}(z) = E(z)$ and $\eta H_{fd}(z) = \eta H(z)$.

For $\alpha = 1$, $E_{fd}(z) = \eta H(z)$ and $\eta H_{fd}(z) = -E(z)$, which is consistent with electromagnetic principle of duality. For any value in between solutions may be regarded as intermediate between the original and dual to the original solutions.

The example of fractional curl taken above can be extended to reflecting media and that solution will enhance the duality of the reflection coefficients.

4.8 Concluding Comments

This chapter extends the argument about reality of having the concept of fractional order calculus to describe the nuclear reactor and application in electromagnetic theory. With regards to the electromagnetic theory, the application of the fractional curl will see extension of its usage in description of Left Handed Materials (LHM) or Metamaterials, where the electromagnetic gets reversed. Reversal of Snell's law, reversed Doppler effect, and superluminality are terms associated with LHM utilized presently to have perfect focusing of beams by straight surfaces. Future will see the geometrical interpretation of the concept of fractional curl and its use in formation of turbulence in flow of fluids and electromagnetics. The concept of fractional divergence as introduced in reactor description will in future lead to development of reactor criticality concepts based on fractional geometrical buckling. This enables to describe the reactor flux profile more closely to actual and maintain efficient correction and control. The fractional divergence will be used to describe several anomalous effects presently observed in diffusion experiments, which is presently ode to non-linearity effects and its explanation through integer order theory or by probabilistic methods.

Chapter 5
Fractional Differintegrations: Insight Concepts

5.1 Introduction

This chapter describes the geometric and physical interpretation of fractional integration and fractional differentiation. As a start point, the Reimann–Liouville (RL) fractional integration is taken. The geometric interpretation is developed first for the RL integration process along with the concept of transformed time scales, and in-homogeneous time axis. Thereafter the RL definition is geometrically explained by convolution of the power function and the integrand. The concept of delay is developed for Grunwald–Letnikov differintegration process, and this is converted into the specific definition of short-memory principle, used for computer applications. The GL differintegration is also explained as in the classical calculus by considering infinitesimal quantities for the independent variable and the function, and explained graphically. The GL definition is expanded with binomial coefficients and its application to numerical regression. Here the concept of generating function is discussed, which by either power series expansion or continued fraction expansion approximates fractional operators. These methods are advance algorithms to get digital realization for fractional order controllers. Small introduction is made regarding the definitions of local fractional derivatives for continuous but nowhere differentiable functions.

5.2 Symbol Standardization and Description for Differintegration

Mathematicians have used several notations since the birth of fractional calculus. Several contemporary notations for fractional differentiation and fractional integration are mentioned in Sect. 1.4. Here, attempt will be made to standardize the notations as differintegrals. The same operator is used as integrator, when index is negative, and differentiator, when index is positive. Separate notation will be used to indicate initialized differintegral operator and un-initialized operator. However, the difference in notations is made clear as "un-initialized" and "initialized" differintegrals; the concept of initialization function $\psi(f, q, a, c, t)$ is dealt in detail in the next chapters 6 and 7.

$_cD_t^q f(t)$ represents the initialized qth order differintegration of $f(t)$ from start point c to t. $_cd_t^q f(t)$ represents un-initialized generalized (or fractional) qth order differintegral. This is also same as

$$\frac{d^q f(t)}{[d(t-c)]^q} \equiv {_cd_t^q} f(t),$$

shifting the origin of function at start of the point from where differintegration starts. This un-initialized operator can also be shortened to the form $d^q f(t)$. The initialization function (not a constant) is represented as $\psi(f, q, a, c, t)$, meaning that this is function of independent variable t, and is for differintegral operator of order q, for the function $f(t)$ born at $t = a$ (before that the function is zero), and differintegral process starting at $t = c$. This initialization function can be short formed as $\psi(t)$, $\psi(f, q, t)$. Therefore the expression between initialized differintegral and un-initialized one is

$$_cD_t^q f(t) = {_cd_t^q} f(t) + \psi(f, q, a, c, t)$$

The notation contains lower limit of the process at the front subscript and the order of the process at the tail superscript, with independent variable with respect to what is being differintegrated.

5.3 Reimann–Liouville Fractional Differintegral

5.3.1 Scale Transformation

The integration in fractional calculus is the embedded part of the fractional differintegration. The RL definition is described as follows:

$$_0D_t^{-\alpha} f(t) = \frac{1}{\Gamma(\alpha)} \int_0^t f(\tau)(t-\tau)^{\alpha-1} d\tau$$

Take the function g as which is basically scaling the time τ variable to the function $g(\tau)$, it is the scale transformation concept. It is described as follows:

$$g(\tau) = \frac{1}{\Gamma(\alpha+1)} \{t^\alpha - (t-\tau)^\alpha\}$$

$$\tau \rightarrow g(\tau)$$

$$dg(\tau) = \frac{(t-\tau)^{\alpha-1}}{\Gamma(\alpha)}$$

5.3 Reimann–Liouville Fractional Differintegral

$$_0D_t^{-\alpha}f(t) = \int_0^t f(\tau)dg(\tau)$$

Therefore, the fractional integration of the function is area under the curve for the plot of $f(\tau)$ and $g(\tau)$, from 0 to t. Let us take three axes $\tau, g(\tau), f(\tau)$, making a cubic room with floor comprising of plane τ and $g(\tau)$. We plot the function as from $0 < t < \tau$, in the floor

$$g(\tau) = \frac{1}{\Gamma(\alpha+1)}\{t^\alpha - (t-\tau)^\alpha\}.$$

This is depicted in Fig. 5.1. Along the obtained curve (on the floor), we build a fence of varying height $f(\tau)$, so the top edge of the fence is a three-dimensional line. Points are $\tau, g(\tau), f(\tau)$ for $0 < \tau < t$.

The shadow on the wall $(\tau, f(\tau))$ as shown in Fig. 5.1 is a well-known area under the curve and is a normal integer order integration $_0D_t^{-1}f(t) = \int_0^t f(\tau)d\tau$. The second shadow on the wall $(g(\tau), f(\tau))$ is geometric interpretation of the fractional integration of $f(t)$, i.e. $_0D_t^{-\alpha}f(t)$ for fixed t. The observation from Fig. 5.1 is that what happens when t is changing (namely growing), the fence changes simultaneously. Its length and in certain sense its shape changes.

The wall $(f(\tau), g(\tau))$, depicting shadow growing as $t = 9 \to 10$ depicting fractional integration, is shown in Fig. 5.2.

For $t = 10$, τ is varied from 0 to 10. $g(\tau)$ is formed and then $f(\tau)$ is plotted. Figure 5.2 shows the change in the shape of the curve from $t = 9$ to $t = 10$ and the integration under the new shape. The difference in the integer order integration

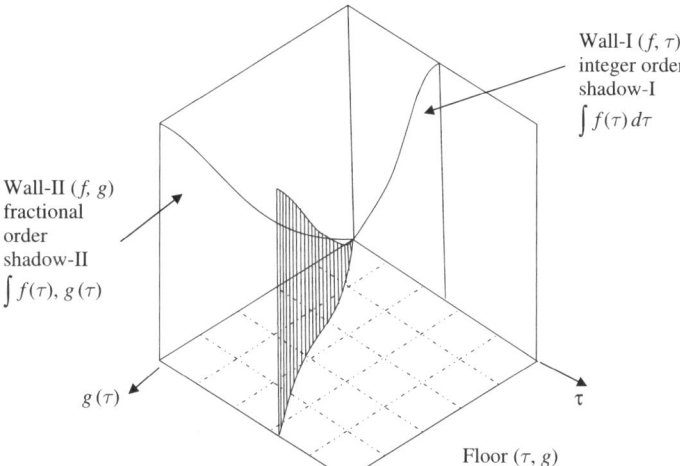

Fig. 5.1 The three-dimensional representation of RL fractional integration

Fig. 5.2 Shadow on the wall showing fractional integration as t grows

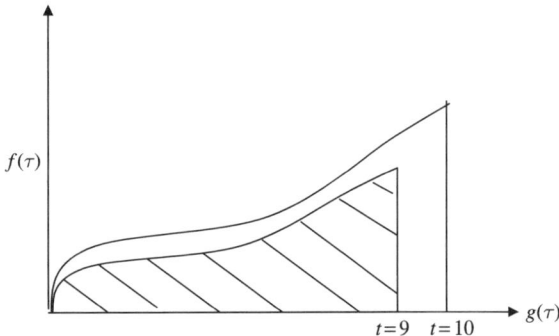

is that of the new shape of the curve from 0 to 10 as compared to old one 0–9. When t changes, the entire preceding time interval changes as well. Let us consider a moving object fitted with speedometer and a clock. The observer in the moving object records the speed at the end of each second (time interval) as $v_1, v_2, v_3, \ldots v_{10}$. At the end of 10 s, the observer in the moving object calculates the distance, $S_N = v_1 + v_2 + v_3 + \ldots v_{10}$, where the observer takes by the local clock each interval of time as "one-second." Now the observer stationary at the fixed frame of reference knows that the clock of the mobile observer is running slow. The actual time at the end of each one-second is 1, 2, 4, 8, 16, 32, 64..... Then the actual distance traveled, as seen by stationary observer (call cosmic observer) as per the actual time (call cosmic time), is $S_o = v_1 + 2v_2 + 4v_3 + 8v_4 + 16v_5 + 32v_4 + 64v_5 + \ldots$, is much more than S_N. The integration of local velocity (speed) with local time is integer order integration, and the integration of the speed recorded with respect to the transformed time is fractional order integration. When moving body changes its position in space–time, the gravitation field in the entire space–time changes due to movement of the object. As a consequence, the "cosmic time interval", which corresponds to the history of the movement of the moving object, changes. S_N, $v(\tau)$, τ is individual, distance, speed, and time, of the observer moving with the object. S_o, $v_o(t)$, t, are the distance, speed, and time as recorded by the observer outside on fixed frame of reference (cosmic). They are related as

$$S_N(t) = {}_0D_t^{-1}v(t) = \int_0^t v(\tau)d\tau, \; S_o(t) = {}_0D_t^{-\alpha}v(t) = \int_0^t v(\tau)dg(\tau), \; v(t) = {}_0D_t^{-\alpha}S_o(t)$$

is individual velocity of the local speedometer as related to cosmic distance, and $v_o(t) = \frac{d}{dt}S_o(t) = \frac{d}{dt}{}_0D_t^{-\alpha}v(t) = {}_0D_t^{1-\alpha}v(t)$. The first derivative of cosmic distance is the cosmic velocity, and the cosmic velocity is fractional derivative of order $(1-\alpha)$ of the local velocity. Figure 5.3 gives the two kind of time: homogeneous (local time) and heterogeneous (transformed time). In the above example, the variable t is used as a notation for both observers, N and O.

$g(\tau)$ describes the in-homogeneous time scales, which depends not only on τ but also on the parameter t, representing the last measured value of the individual time

5.3 Reimann–Liouville Fractional Differintegral

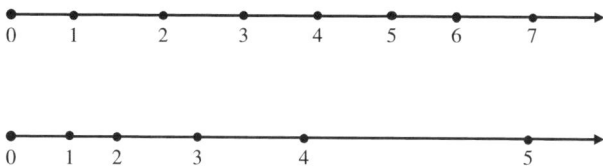

Fig. 5.3 Homogeneous and heterogeneous time

(of the moving object). Fractional integration in time means transformation of the local time to cosmic time.

5.3.2 Convolution

The Reimann–Liouville definition for the fractional integral is

$$\frac{d^{-q}}{[d(t-a)]^{-q}} f(t) = \frac{1}{\Gamma(q)} \int_a^t (t-\tau)^{q-1} f(\tau) d\tau, \quad q \geq 0$$

This definition is extended to fractional differentiation with m as an integer, as follows:

$$\frac{d^{m-q}}{[d(t-a)]^{m-q}} f(t) \equiv \frac{1}{\Gamma(q)} \frac{d^m}{dt^m} \int_a^t (t-\tau)^{q-1} f(\tau) d\tau, \quad q \geq 0, m > q$$

The RL integral can be viewed as convolution of the integrand function with power function; when both the functions are causal, it can be expressed as

$$\frac{d^{-q}}{[d(t-0)]^{-q}} f(t) = f(t)*h(t) = f(t)* \left(\frac{1}{\Gamma(q)t^{-q+1}} \right) = \frac{1}{\Gamma(q)} \int_0^t \frac{f(\tau)d\tau}{(t-\tau)^{-q+1}}$$

where

$$h(t) = \frac{1}{\Gamma(q)t^{-q+1}}.$$

Causal functions mean that no convolution response can be obtained before the function $f(t)$ is applied Fig. 5.4 demonstrates the convolution process. The function is

$$f(t) = \cos\left(\frac{2\pi}{5}t\right);$$

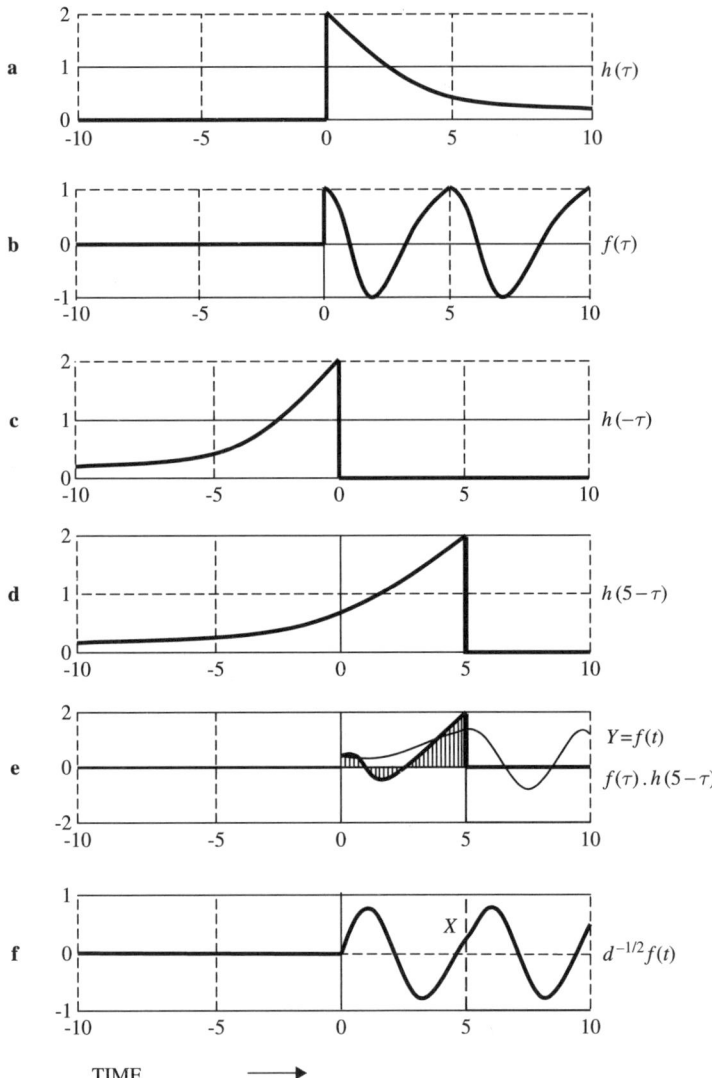

Fig. 5.4 RL integral interpreted as convolution

the order of fractional integration is half $q = 1/2$. Figure 5.4a shows $h(\tau)$ versus τ; Fig. 5.4b shows that $f(t), t > 0$. Figure 5.4c shows $h(-\tau)$. The curve is obtained for the value $t = 5$, and Fig. 5.4d shows the plot of $h(t - \tau)$ at $t = 5$, i.e., $h(5 - \tau)$ versus τ. Figure 5.4e shows the full integrand for $t = 5$. Now moving this $h(t - \tau)$ for several continuous values of t from 0 to 10, repeating the graphs **d** and **e** and obtaining the value of the integral of the product (for several values of t) the final graph **f** is obtained. For $t = 5$, the graph **e** shows full integrand as $h(5 - \tau)f(\tau)$. The integral of this product becomes $t = 5$, value of the semi-integral of $f(t)$.

Fig. 5.5 Convoluting function $h(t)$ for several t

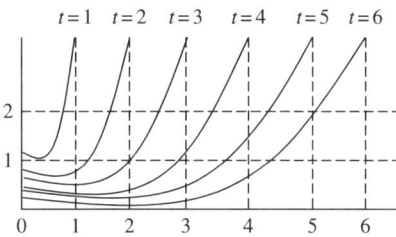

In Fig. 5.4f, the point X is

$$\int_0^5 f(\tau)h(5-\tau)d\tau,$$

definite value of the integration.

Figure 5.5 demonstrates the several $h(t-\tau)$ for $t = 1, 2, 3, 4, 5, 6, \ldots$

Figure 5.6 demonstrates the semi-derivative of

$$f(t) = \cos\left(\frac{2\pi}{5}t\right),$$

which is obtained from differentiating once the RL semi-integral graph.

5.3.3 Practical Example of RL Differintegration in Electrical Circuit Element Description

These types of intermediate devises are becoming reality in electrical circuits as evident from patent US 20060-267595 of November 2006. We shall start with a resistoductance alone, which is a linear circuit element whose behavior is intermediate between that of an inductor element and ohmic resistor element. The term

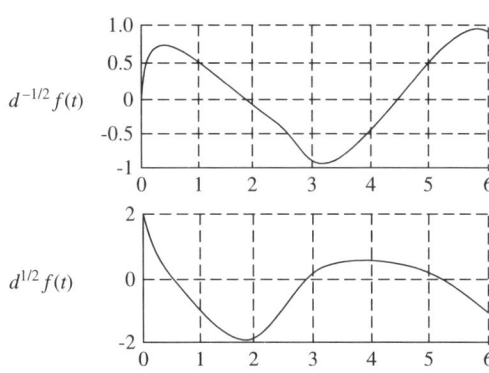

Fig. 5.6 Semi-derivative of function. Differentiating once the RL semi-integral of $f(t)$

resistoductance is combination of pure resistance and pure inductor. As integer order equations, the fractional order requires fractional derivatives (or integrals) as initial conditions. How does then one relate to initial condition expressed in terms of fractional differintegrals? The constitutive equation of such an element is

$$v(t) = K_0 D_t^\alpha i(t) \text{ or } i(t) = \frac{1}{K} {}_0 D_i^{-\alpha} v(t)$$

Here $v(t)$ is across variable, i.e., the voltage across the circuit element, and $i(t)$ is the through variable, i.e., current through the circuit element. If $\alpha = 0$, the circuit is purely resistive and $K = R$ ohms, and if $\alpha = 1$, the circuit is purely inductive and $K = L$ henrys.

For a step input of voltage $v(t) = V_0$ applied at $t = 0$, the current is described as fractional integration of the forcing function, i.e., $i(t) = \frac{1}{K} {}_0 D_t^{-\alpha} v(t)$. In terms of convolution definition (as forcing function being causal) and $h(t) = \frac{t^{\alpha-1}}{\Gamma(\alpha)}$, the power function, the current is obtained as follows:

$$i(t) = \frac{1}{K}[h(t)*v(t)] = \int_0^t \frac{(t-\tau)^{\alpha-1}}{K\Gamma(\alpha)} V_0 d\tau = \frac{V_0}{K\Gamma(\alpha)} \int_t^0 -(x)^{\alpha-1} dx$$

$$= \frac{V_0}{K\Gamma(\alpha)} \left[-\frac{x^{\alpha-1+1}}{\alpha-1+1} \right]_t^0 = \frac{V_0}{K\Gamma(\alpha)} \left[-\frac{x^\alpha}{\alpha} \right]_t^0 = \frac{V_0}{K\alpha\Gamma(\alpha)} t^\alpha$$

Using $\Gamma(\alpha + 1) = \alpha\Gamma(\alpha)$, we obtain the current to step input as

$$i(t) = \frac{V_0}{K\Gamma(\alpha+1)} t^\alpha$$

for pure resistance $\alpha = 0$, and $i(t) = V_0/K$ and for pure inductance $\alpha = 1$, and $i(t) = (V_0/K)t$.

The initial value of the current vanishes, i.e., there is no instantaneous current, only retarded response. However, the first ordinary derivative of $i(t) = K_1 t^\alpha$ is unbounded, so that a finite though undefined current can be reached in arbitrary small time interval. The change of $i(t)$ is described by the fractional differential equation as ${}_0 D_t^\alpha i(t) = V_0/K$.

In accordance with the theory of fractional differential in terms of RL derivatives, an initial condition involving ${}_0 D_t^{\alpha-1} i(t)$ is thus required. Physically this initial condition has no representation and cannot be directly obtained from measurement. This condition can be found by taking the first-order integral of the constitutive equation. This process relates the fractional (immeasurable) initial condition to something of reality and measurable as

$$\left[{}_0 D_t^{\alpha-1} i(t) \right]_{t\to 0} = \left[{}_0 D_t^{-1} (V_0/K) \right]_{t\to 0}$$

5.3 Reimann–Liouville Fractional Differintegral

In the case under consideration, voltage stress is finite at all times hence $\left[{}_0D_t^{-1}V_0\right]_{t\to 0} = 0$, which leads to the condition of zero initial condition involving fractional differintegral, namely, $\left[{}_0D_t^{\alpha-1}i(t)\right]_{t\to 0} = 0$. The same consideration applies to general finite voltage $v(t)$, and the equation to be solved is ${}_0D_t^{\alpha}i(t) = v(t)/K$, and same zero initial condition be attached.

Now for the impulse input voltage at time 0, i.e., $v(t) = B\delta(t)$ at $t = 0$, the current expression is again $i(t) = \frac{1}{K}{}_0D_t^{-\alpha}v(t)$ using the convolution definition and $h(t) = \frac{t^{\alpha-1}}{\Gamma(\alpha)}$, the current expression is

$$i(t) = \frac{1}{K}\left[h(t)^*v(t)\right] = \frac{1}{K}\left[h(t)^*B\delta(t)\right] = \frac{B}{K}h(t) = B\frac{t^{\alpha-1}}{K\Gamma(\alpha)}$$

This is obtained as the convolution of function with impulse at $t = 0$, this returns the function itself. The power function with gain B is retuned. This is property of the convolution. As observed from the derived current expression for the impulse voltage, the initial voltage–stress singularity gives rise to lower order current singularity, since resistoductance cannot respond instantaneously.

The impulse response is mathematical convenience to evaluate transfer characteristics and is seldom used in practice because it is even more problematic to apply homogeneous impulse voltage on circuit element than to apply step. However investigating the impulse response will follow the same reasoning as for the step.

For the impulse voltage excitation for $t > 0$, the fractional differential equation is ${}_0D_t^{\alpha}i(t) = 0$. In accordance with the theory of fractional differential equations with RL derivatives, an initial condition involving $[{}_0D_t^{\alpha-1}i(t)]_{t\to 0}$ is required. This can be found through integration of constitutive equation as

$$\left[{}_0D_t^{\alpha-1}i(t)\right]_{t\to 0} = \left[{}_0D_t^{-1}(v(t)/K)\right]_{t\to 0} = B/K$$

which gives the initial condition in terms of fractional differintegral as $\left[{}_0D_t^{\alpha-1}i(t)\right]_{t\to 0} = B/K$. This fractional differintegral initial condition is non-zero, well defined, and bounded, whereas both current and its integer order derivatives are unbounded, and its first-order integral is zero so that a meaningful initial condition expressing the loading conditions cannot be obtained using integer order derivatives.

In the above example of resistoductance, it is possible to attribute physical meaning to initial condition expressed in terms of fractional differintegral. Expressing initial condition in terms of fractional derivative of a function $u(t)$ is not a problem because it does not require a direct experimental evaluation of these fractional derivatives. Instead, one should consider its counterpart (in separable twin), $v(t)$ via basic physical law, and measure (or consider) its initial values.

Similarly, other intermediate models can be considered as resistocaptance. Resistocaptance will be similar in nature to resistoductance, where the circuit element will be intermediate between pure resistance and pure capacitance. The constitutive part will be

$$v(t) = \frac{1}{K} {}_0D_t^{-\alpha} i(t) \text{ or } i(t) = K {}_0D_t^{\alpha} v(t),$$

where for $\alpha = 1$ the element is pure capacitor with $K = C$ farads, and for $\alpha = 0$ the element will be pure conductance $K = G$ mho.

These intermediate models can explain the behavior of "time-constant" dispersion effects in the circuit behavior when the relaxation observations cannot be explained by single $\tau = R/L$ or RC time constant.

5.4 Grunwald–Letnikov Fractional Differinteration

The basic definition of Grunwald–Letnikov (GL) is

$$\frac{d^q f(t)}{[d(t-a)]^q} \equiv \lim_{N \to \infty} \frac{\left(\frac{t-a}{N}\right)^{-q}}{\Gamma(-q)} \sum_{j=0}^{N-1} \frac{\Gamma(j-q)}{\Gamma(j+1)} f\left(t - j\left[\frac{t-a}{N}\right]\right)$$

When the index q is negative, the above process is fractional integration, and when q is positive fraction, the process tends to fractional differentiation. For understanding the recursive formulation of GL method consider a small example with $q = 1/2$, $N = 4$. The GL expansion for the four terms is

$$\frac{d^{1/2} f(t)}{[d(t-a)]^{1/2}} \equiv \frac{\left(\frac{t-a}{N}\right)^{-1/2}}{\Gamma\left(-\frac{1}{2}\right)} \left\{ \begin{array}{l} \frac{\Gamma(-1/2)}{\Gamma(1)} f(t) + \\ \frac{\Gamma(1/2)}{\Gamma(2)} f\left(t - \left(\frac{t-a}{4}\right)\right) + \\ \frac{\Gamma(3/2)}{\Gamma(3)} f\left(t - 2\left(\frac{t-a}{4}\right)\right) + \\ \frac{\Gamma(5/2)}{\Gamma(4)} f\left(t - 3\left(\frac{t-a}{4}\right)\right) \end{array} \right\}$$

The process explanation is that the function $f(t)$ is first multiplied by a constant then time shifted by the amount $((t-a)/N)$, for $(N-1)$ many times, and each shifted term gets a weight multiplication $(\Gamma(j-q)/\Gamma(-q)\Gamma(j+1))$ before summing and then scaled by a factor $((t-a)/N)^{-q}$. Figure 5.7 shows the diagrammatic representation of weighted addition for the four terms as derived for semi-differentiation of the function $f(t)$.

Thus the observation is that semi-derivative evaluation for a function by GL method is seen to be a summation of progressively delayed evaluation of $f(t)$ multiplied by progressively decreasing constant weights, and finally multiplied by $((t-a)/N)^{-q}$. In reality choosing the value $N = 4$ is a crude approximation. The definition states $N \to \infty$, so take a large value, as $N = 10,000$. The following case is considered for $q = 1/2$, $N = 10,000$:

5.4 Grunwald–Letnikov Fractional Differinteration

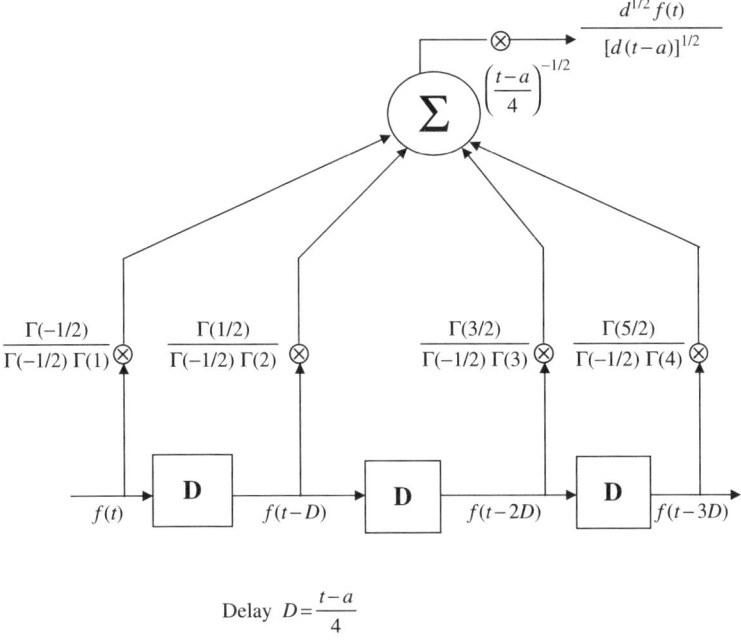

Fig. 5.7 Time delay weighted summation of semi-derivative (four-terms)

$$\frac{d^{1/2} f(t)}{[d(t-a)]^{1/2}} \equiv \frac{\left(\frac{t-a}{10,000}\right)^{-1/2}}{\Gamma\left(-\frac{1}{2}\right)} \left\{ \begin{array}{l} \frac{\Gamma(-1/2)}{\Gamma(1)} f(t) + \\ \frac{\Gamma(1/2)}{\Gamma(2)} f\left(t - \left(\frac{t-a}{10,000}\right)\right) + \\ \frac{\Gamma(3/2)}{\Gamma(3)} f\left(t - 2\left(\frac{t-a}{10,000}\right)\right) + \\ * \\ \frac{\Gamma(j-1/2)}{\Gamma(j+1)} f\left(t - j\left(\frac{t-a}{10,000}\right)\right) + \\ * \\ \frac{\Gamma(9997.5)}{\Gamma(9999)} f\left(t - 9998\left(\frac{t-a}{10,000}\right)\right) + \\ \frac{\Gamma(9998.5)}{\Gamma(10,000)} f\left(t - 9999\left(\frac{t-a}{10,000}\right)\right) \end{array} \right\}$$

By comparison, it is seen that the four-term expansion and the 10,000-term expansion of the semi-derivative in GL method Gamma function based coefficients are the same for the first N terms. The time shift factor (incremental delay) is of course very much smaller and indeed approaches zero as the number of terms are made toward infinity.

Figure 5.8 shows the higher order approximation with large number of terms. Figures 5.7 and 5.8 are similar except that the delay steps are smaller in the higher order approximation. One observation is made as the time (independent variable) grows, the delay between the consecutive terms increases as $\Delta T = \frac{t-a}{N}$ is the delay.

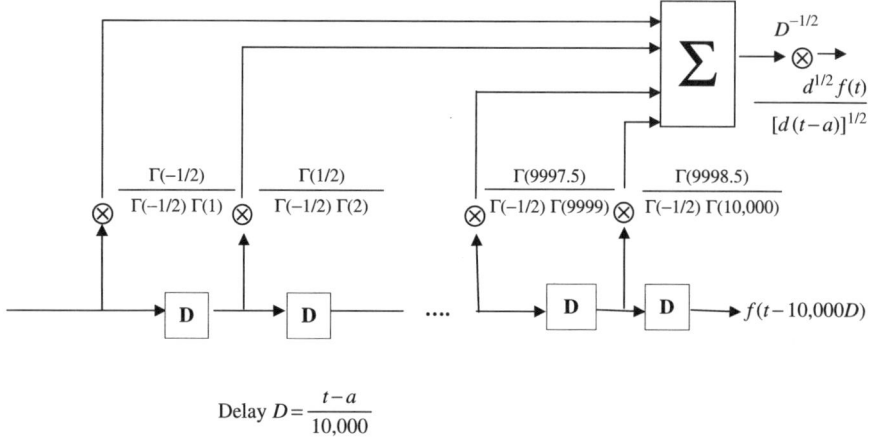

Fig. 5.8 Time delay weighted summation of semi-derivative (large number terms) higher order approximation of GL

Meaning that for fixed N, as the time grows the resolution between the samples of the function decreases. This is overcome by spreading the sampling instrument N uniformly in number in one interval and thus keeping ΔT same the GL approximate can be written as

$$_aD_t^q f(t) = \lim_{\Delta T \to 0} \frac{(\Delta T)^{-q}}{\Gamma(-q)} \sum_{j=0}^{N-1} \frac{\Gamma(j-q)}{\Gamma(j+1)} f(t - j(\Delta T))$$

5.5 Unification of Differintegration Through Binomial Coefficients

The coefficient in Figs. 5.7 and 5.8 for the GL approximation in series form however is through Gamma functions. Practically it is tough to evaluate Gamma function especially when the argument is a large number. Also for electronics and real time computer realization, the use of Gamma function is tough. Here the unification approach is presented by use of binomial coefficients. First integer order unification of repeated differentiation and repeated folded integration is presented. The same can be generalized by use of Letnikov theorem to any arbitrary order (not in the scope here).

The repeated integer order differentiation is as follows:

$$f^{(1)}(t) = \frac{df}{dt} = \lim_{h \to 0} \frac{f(t) - f(t-h)}{h}$$

5.5 Unification of Differintegration Through Binomial Coefficients

$$f^{(2)}(t) = \frac{d^2 f}{dt^2} = \lim_{h \to 0} \frac{f^{(1)}(t) - f^{(1)}(t-h)}{h}$$

$$= \lim_{h \to 0} \frac{1}{h} \left\{ \frac{f(t) - f(t-h)}{h} - \frac{f(t-h) - f(t-2h)}{h} \right\}$$

$$= \lim_{h \to 0} \frac{f(t) - 2f(t-h) + f(t-2h)}{h^2}$$

$$f^{(3)} = \frac{d^3 f}{dt^3} = \lim_{h \to 0} \frac{f(t) - 3f(t-h) + 3f(t-2h) - f(t-3h)}{h^3}$$

By induction,

$$f^{(n)} = \frac{d^n f}{dt^n} = \lim_{h \to 0} \frac{1}{h^n} \sum_{r=0}^{n} (-1)^r \binom{n}{r} f(t - rh)$$

The binomial coefficient

$$\binom{n}{r} = \frac{n(n-1)(n-2)\ldots(n-r+1)}{r!},$$

for $r > 1$. The above derivative expression is generalized for any p integer where $p \leq n$, is

$$f_h^{(p)}(t) = \frac{1}{h^p} \sum_{r=0}^{n} (-1)^r \binom{p}{r} f(t - rh)$$

and obviously

$$\lim_{h \to 0} f_h^{(p)}(t) = f^{(p)}(t) = \frac{d^p f}{dt^p}$$

because in such a case as follows, all the coefficients after $\binom{p}{p}$ are zero.

For integration case, the index is negative; for convenience sake let us define (like binomial coefficient) the following:

$$\left[\begin{matrix} p \\ r \end{matrix} \right] = \frac{p(p+1)(p+2)\ldots(p+r-1)}{r!}.$$

Call this as pseudo-binomial coefficient.
Then

$$\binom{-p}{r} = \frac{-p(-p-1)(-p-2)\ldots(-p-r+1)}{r!} = (-1)^r \left[\begin{matrix} p \\ r \end{matrix} \right].$$

Using this, we have for integration (differentiation with negative index):

$$f_h^{(-p)}(t) = \frac{1}{h^{-p}} \sum_{r=0}^{n} \begin{bmatrix} p \\ r \end{bmatrix} f(t-rh) = h^p \sum_{r=0}^{n} \begin{bmatrix} p \\ r \end{bmatrix} f(t-rh)$$

Taking interval $t - a$ and dividing them equally by n, the limit can be written as follows:

$$\lim_{\substack{h \to 0 \\ nh = t-a}} f_h^{(-p)} = {}_aD_t^{(-p)} f(t)$$

Take $p = 1$, then $f_h^{(-1)}(t) = h \sum_{r=0}^{n} f(t-rh)$. Taking into account $t - nh = a$,

$$\lim_{\substack{h \to 0 \\ nh = t-a}} f_h^{(-1)}(t) = {}_aD_t^{-1} f(t) = \int_0^{t-a} f(t-z)dz = \int_a^t f(\tau)d\tau$$

Take $p = 2$, then $\begin{bmatrix} 2 \\ r \end{bmatrix} = \frac{2 \cdot 3 \cdot 4 \ldots (2+r-1)}{r!} = r + 1$, giving

$$f_h^{(-2)}(t) = h^2 \sum_{r=0}^{n} (r+1) f(t-rh).$$

Rearranging and putting $t + h = y$, we get

$$f_h^{(-2)}(t) = h \sum_{r=0}^{n} [(rh)f(t-rh) + f(t-rh)] = h \sum_{r=1}^{n+1} (rh)f(y-rh)$$

and taking $h \to 0$ we get

$$\lim_{\substack{h \to 0 \\ nh = t-a}} f_h^{(-2)}(t) = {}_aD_t^{(-2)} f(t) = \int_0^{t-a} z f(t-z)dz = \int_a^t (t-\tau)f(\tau)d\tau.$$

Take $p = 3$, then $\begin{bmatrix} 3 \\ r \end{bmatrix} = \frac{3 \cdot 4 \ldots (3+r-1)}{r!} = \frac{(r+1)(r+2)}{1 \cdot 2}$ we have

5.6 Short Memory Principle: A Moving Start Point Approximation and Its Error

$$f_h^{(-3)}(t) = \frac{h}{1.2} \sum_{r=0}^{n} (r+1)(r+2)h^2 f(t-rh) \to f_h^{(-3)}(t)$$

$$= \frac{h}{1.2} \sum_{r=1}^{n+1} r(r+1)h^2 f(y-rh),$$

here also $t + h = y$ is substituted. Expressing the above by rearranging:

$$f_h^{(-3)}(t) = \frac{h}{1.2} \sum_{r=1}^{n+1} (rh)^2 f(y-rh) + \frac{h^2}{1.2} \sum_{r=1}^{n+1} (rh) f(y-rh),$$ taking $h \to 0$ we

obtain $_aD_t^{(-3)} f(t) = \frac{1}{2!} \int_0^{t-a} z^2 f(t-z) dz = \int_a^t (t-\tau)^2 f(\tau) d\tau$. Because $y \to t$, as $h \to 0$ and

$$\lim_{\substack{h \to 0 \\ nh = t-a}} \frac{h^2}{1.2} \sum_{r=1}^{n+1} rh f(y-rh) = \lim_{\substack{h \to 0 \\ nh = t-a}} h \int_a^t (t-\tau) f(\tau) = 0$$

Generally this process suggests that

$$_aD_t^{-p} f(t) = \lim_{\substack{h \to 0 \\ nh = t-a}} h^p \sum_{r=0}^{n} \begin{bmatrix} p \\ r \end{bmatrix} f(t-rh) = \frac{1}{(p-1)!} \int_a^t (t-\tau)^{p-1} f(\tau) d\tau,$$

repeated p-fold integration.

Applying the Letnikov theorem, the above expressions can be generalized for any arbitrary order of differentiation and integration process with binomial coefficients expressed as weights as indicated below.

For arbitrary differentiation of real order, the coefficient approximation for $(-1)^k \begin{pmatrix} \alpha \\ k \end{pmatrix}$ is $\omega_0^{(\alpha)} = 1$ and $\omega_k^{(\alpha)} = \left(1 - \frac{1+\alpha}{k}\right) \omega_{k-1}^{(\alpha)}$; it is used for recursive computation.

For arbitrary integration of real order, the coefficient approximation for $\begin{bmatrix} \alpha \\ k \end{bmatrix}$ is $\omega_0^{(-\alpha)} = 1$ and $\omega_k^{(-\alpha)} = \left(1 - \frac{1-\alpha}{k}\right) \omega_{k-1}^{(-\alpha)}$; it is used for recursive computation.

A caution is put here; the factorials in the binomial (for differentiation) and pseudo-binomial (for integration) coefficients get generalized by Gamma functions as indicated in Chap. 1, for arbitrary order. However these weights are approximates and are helpful in recursive formulation for computation and real time applications, obviously with error.

5.6 Short Memory Principle: A Moving Start Point Approximation and Its Error

For $t \gg a$, the number of addends in fractional differintegral approximates becomes enormously large. However, it follows from the expressions for GL definitions and unification arguments in the preceding sections that for large t the role of "history" of the behavior of the function $f(t)$ near the lower terminal $t = a$ (the start point of the differintegral) process can be neglected under certain assumptions. Those observations lead us to the formulation of the "short-memory" principle, which means taking into account only the "recent past" of the function behavior, that is, in the interval $[t - L, t]$, where L is memory length (in unit of time). Therefore the approximation is

$$_a D_t^\alpha f(t) \approx {}_{t-L} D_t^\alpha f(t), \ (t > a + L)$$

In other words, according to the short-memory principle, the fractional differintegrational with lower limit a is approximated by fractional differintegration with "moving lower limit" $(t - L)$. With this approximation, the addends in the GL process is always not greater than $[L/h]$. In the selection of number of addend, the rule followed is thus by this short-memory principle (for $N(t)$) is $N(t) = \min\left\{\left[\frac{t}{h}\right], \left[\frac{L}{h}\right]\right\}$, h is step size.

Of course, for this simplification, penalty is paid in terms of accuracy. Following rule will explain this. We consider the function in the interval (a, b) to be bounded by $f(t) \leq M$ and have the error value as per required accuracy as ε. Then, the following estimate is the rule:

Error in short-memory principle is expressed as

$$\Delta(t) = \left|_a D_t^\alpha f(t) - {}_{t-L} D_t^\alpha f(t)\right| \leq \frac{ML^{-1}}{|\Gamma(1-\alpha)|}$$

where $(a + L \leq t \leq b)$ and $f(t) \leq M$ for $a \leq t \leq b$.

This inequality rule can be used for determining the "memory-length" (in unit of time), provided the required accuracy is met, i.e., $\Delta(t) \leq \varepsilon$ and $a + L \leq t \leq b$, therefore

$$L \geq \left(\frac{M}{\varepsilon |\Gamma(1-\alpha)|}\right)^{1/\alpha}$$

Summarizing the approximates for differintegration process, we can write

$$_a D_t^\alpha = \begin{cases} \frac{d^\alpha}{dt^\alpha} & \Re e(\alpha) > 0 \\ 1 & \Re e(\alpha) = 0 \\ \int (d\tau)^{-\alpha} & \Re e(\alpha) < 0 \end{cases}$$

$$_{(t-L)} D_t^{\pm\alpha} f(t) \approx h^{-(\pm\alpha)} \sum_{j=0}^{N(t)} \omega_j^{\pm\alpha} f(t - jh)$$

$$N(t) = \min\left\{\left[\frac{t}{h}\right], \left[\frac{L}{h}\right]\right\}$$

$$\omega_0^{\pm\alpha} = 1, \ \omega_j^{\pm\alpha} = \left(1 - \frac{1 \pm \alpha}{j}\right) \omega_{j-1}^{\pm\alpha}$$

5.7 Matrix Approach to Discretize Fractional Differintegration and Weights

The weights or the coefficients for approximation of fractional differintegration as described in the preceding sections are the following:

For differentiation: $\omega_k^\alpha = (-1)^k \binom{\alpha}{k}$, for $k = 0, 1, 2, 3 \ldots N$.

This can be written as

$$\omega_0^\alpha = 1, \omega_k^\alpha = \left(1 - \frac{1+\alpha}{k}\right), \text{ for } k = 1, 2, 3 \ldots N.$$

For integration: $\omega_k^{-\alpha} = \binom{\alpha}{k}$, for $k = 0, 1, 2, 3, \ldots N$.

This can be written as

$$\omega_0^{-\alpha} = 1, \omega_k^{-\alpha} = \left(1 - \frac{1-\alpha}{k}\right), \text{ for } k = 1, 2, 3, ..N.$$

Both are same formulation. However, in system identification, the most appropriate value of order α must be found; this means that various values of α are considered, and for each α, the we have to calculate ω_k^α. In such cases, the above recursive method is not easy; instead, Fast Fourier Transform (FFT) is used. The weights ω_k^α can be considered (for differentiation) as coefficient of the series function $(1 - z)^\alpha$. The power series is expanded as

$$(1-z)^\alpha = \sum_{k=0}^{\infty} (-1)^k \binom{\alpha}{k} z^k = \sum_{k=0}^{\infty} \omega_k^{(\alpha)} z^k$$

Substituting $z = e^{-j\varphi}$, we have $(1 - e^{-j\varphi})^\alpha = \sum_{k=0}^{\infty} \omega_k^{(\alpha)} e^{-j\varphi}$ and the coefficient $\omega_k^{(\alpha)}$ is expressed in FFT as $\omega_k^{(\alpha)} = \frac{1}{j2\pi} \int_0^{2\pi} f_\alpha(\varphi) e^{jk\varphi} d\varphi$, where $f_\alpha(\varphi) = (1 - e^{-j\varphi})^\alpha$.

$\omega_k^{(\alpha)}$ can be computed using FFT. Since in this case we always obtain a finite number of coefficients, the FFT can be used with short-memory principle.

The approximation to fractional derivative can be written as the fractional difference approach:

$$_a D_{t_k}^\alpha f(t) \approx \frac{\Delta^\alpha f(t_k)}{h^\alpha} = h^{-\alpha} \sum_{j=0}^{k} (-1)^j \binom{\alpha}{j} f_{k-j}, \text{ for } k = 1, 2, 3, \ldots N$$

Following the formulation of the above in matrix notation is helpful for coefficient evaluation:

$$\begin{bmatrix} h^{-\alpha}\Delta^{\alpha}f(t_0) \\ h^{-\alpha}\Delta^{\alpha}f(t_1) \\ * \\ h^{-\alpha}\Delta^{\alpha}f(t_{N-1}) \\ h^{-\alpha}\Delta^{\alpha}f(t_N) \end{bmatrix} = \frac{1}{h^{\alpha}} \begin{bmatrix} \omega_0^{(\alpha)} & 0 & 0 & * & 0 \\ \omega_1^{(\alpha)} & \omega_0^{(\alpha)} & 0 & * & 0 \\ * & * & * & 0 & 0 \\ \omega_{N-1}^{(\alpha)} & \omega_{N-2}^{(\alpha)} & * & \omega_0^{(\alpha)} & 0 \\ \omega_N^{(\alpha)} & \omega_{N-1}^{(\alpha)} & * & \omega_1^{(\alpha)} & \omega_0^{(\alpha)} \end{bmatrix} \begin{bmatrix} f_0 \\ f_1 \\ * \\ f_{N-1} \\ f_N \end{bmatrix}$$

In the above coefficient matrix, the coefficients are symbolically represented as: $\omega_j^{(\alpha)} = (-1)^j \binom{\alpha}{j}$, $j = 0, 1, 2, \ldots N$. With the argument as indicated above, for obtaining them from FFT, we have a generating polynomial, whose coefficients will be the triangular matrix with $\omega_j^{(\alpha)}$, with truncation as described by

$$Q(z) = \sum_{k=0}^{\infty} \omega_k^{(\alpha)} z^k \leftrightarrow trunc_N(Q(z)) \stackrel{def}{=} \sum_{k=0}^{N} \omega_k^{(\alpha)} z^k = Q_N(z).$$ We write the matrix notation in short as $[h^{-\alpha}\Delta^{\alpha}f(t_k)] = [B] \cdot [f_k]$, where matrix $[B]$ includes $1/h^{\alpha}$. We call $B_N^{(\alpha)} = [B] = \beta_\alpha(z) = h^{-\alpha}(1-z)^{\alpha}$.

The approximation for fractional integration follows from B_N by doing inverse operation. Define $I_N^{(\alpha)} = (B_N^{(\alpha)})^{-1}$; the generating polynomial representation will be for integration as $I_N^{(\alpha)} \leftrightarrow \varphi_N = trunc_N(\beta_\alpha^{-1}(z)) = trunc_N(h^{\alpha}(1-z)^{-\alpha})$

For integration operation, the coefficient matrix is

$$I_N^{(\alpha)} = h^{\alpha} \begin{bmatrix} \omega_0^{(-\alpha)} & 0 & 0 & 0 & 0 \\ \omega_1^{(-\alpha)} & \omega_0^{(-\alpha)} & 0 & 0 & 0 \\ * & * & \omega_0^{(-\alpha)} & 0 & 0 \\ \omega_{N-1}^{(-\alpha)} & \omega_{N-2}^{(-\alpha)} & * & * & 0 \\ \omega_N^{(-\alpha)} & \omega_{N-1}^{(-\alpha)} & * & \omega_1^{(-\alpha)} & \omega_0^{(-\alpha)} \end{bmatrix}$$

The integral matrix representation in short is $[h^{\alpha}\Delta^{-\alpha}f(t_k)] = [I_N^{(\alpha)}] \cdot [f_k]$, $k = 0, 1..N$.

The weights for integration symbolically are $\omega_j^{(-\alpha)} = (-1)^j \binom{-\alpha}{j} = \begin{bmatrix} \alpha \\ j \end{bmatrix}$, for $j = 0, 1, 2, 3, \ldots N$.

5.8 Infinitesimal Element Geometrical Interpretation of Fractional Differintegrations

The GL definition is described below:

5.8 Infinitesimal Element Geometrical Interpretation of Fractional Differintegrations

$$_aD_t^q f(t) = \lim_{N \to \infty} \frac{\left(\frac{t-a}{N}\right)^{-q}}{\Gamma(-q)} \sum_{j=0}^{N-1} \frac{\Gamma(j-q)}{\Gamma(j+1)} f\left(t - j\left(\frac{t-a}{N}\right)\right)$$

$$= \lim_{\Delta T \to 0} \sum_{j=0}^{N-1} \frac{\Gamma(j-q)}{\Gamma(-q)\Gamma(j+1)} \frac{f(t - j\Delta T)}{\Delta T^q}$$

$$\Delta T = (t-a)/N, \; N \to \infty, \; \Delta T \to 0$$

The nature of the definition may be explored with the jth and $(j+1)$th term. In a general sense, if the terms are additive and $q < 0$, then integration is in effect. If the terms are differenced and $q > 0$, then differentiation is suggested. Then

$$_aD_t^q f(t) = \lim_{\Delta T \to 0} \left\{ \ldots + \frac{\Gamma(j-q)}{\Gamma(-q)\Gamma(j+1)} \frac{f(t - j\Delta T)}{\Delta T^q} + \frac{\Gamma(j+1-q)}{\Gamma(-q)\Gamma(j+2)} \frac{f(t - (j+1)\Delta T)}{\Delta T^q} + \ldots \right\}$$

Dividing throughout by the coefficients of the jth term, and combining the jth and $(j+1)$th term gives

$$_aD_t^q f(t) = \lim_{\Delta T \to 0} \left\{ \ldots + \alpha \left(\frac{f(t - j\Delta T) + \beta f(t - (j+1)\Delta T)}{\Delta T^q} \right) + \ldots \right\},$$

$$j = 1, 2, 3 \ldots \; \beta = \frac{\Gamma(j+1-q)\Gamma(j+1)}{\Gamma(j-q)\Gamma(j+2)}, \; \alpha = \frac{\Gamma(j-q)}{\Gamma(-q)\Gamma(j+1)}$$

5.8.1 Integration

Now when $q = -1$, then $\beta = 1$ and $\alpha = 1$. The GL equation simplifies for j and $j+1$ terms as

$$_aD_t^{-1} f(t) = \lim_{\Delta T \to 0} \left\{ \ldots + \Delta T \left[f(t - j\Delta T) + f(t - (j+1)\Delta T) \right] + \ldots \right\},$$

$$j = 1, 2, 3 \ldots$$

which is conventional integration.

For simplicity, take $q = 1/2$, then

$$_aD_t^{-1/2} f(t)$$

$$= \lim_{\Delta T \to 0} \left\{ \ldots + \alpha \Delta T^{1/2} \left[f(t - j\Delta T) + \beta_{-\frac{1}{2}} f(t - (j+1)\Delta T) \right] + \ldots \right\},$$

$$j = 1, 2, 3 \ldots$$

$$\beta_{-\frac{1}{2}} = \frac{j + \frac{1}{2}}{j + 1}$$

For all j greater than equal to zero, β is positive. The subscripted symbol indicates semi-integration process. For the general case, the expression is

$$_aD_t^{-q}f(t) = \lim_{\Delta T \to 0}\left\{\ldots + \alpha\Delta T^q\left[f(t-j\Delta T) + \beta_{-q}f(t-(j+1)\Delta T\right] + \ldots\right\},$$
$$j = 1, 2, 3\ldots$$

Thus α and β are always positive, when q is between 0 and -1, and the above summation is seen to be the integration process and indeed fractional. A geometric approximation to this integration is given in Fig. 5.9.

If $q = -1$, then $\sum \alpha f(t - j\Delta T)\Delta T$ is an area represented below the curve $\alpha f(t - j\Delta T)$. If $q = -2$ (outside the domain of consideration here, as q is restricted to -1), then $\sum \alpha f(t - j\Delta T)\Delta T^2$ is a volume. Then the series $\sum \alpha f(t - j\Delta T)\Delta T^q$, for $0 \geq q > -1$, may be thought of as "fractional-area." In Fig. 5.9, fractional integral is the area under the $\alpha f(t - j\Delta T)$ curve multiplied by ΔT^{q-1}, a fractional scaled version of the shaded area or full area.

5.8.2 Differentiation

For $q = 1$, $\alpha = 0$ for all j except for $j = 0$. The GL expansion is

$$_aD_t^1 f(t) = \lim_{\Delta T \to 0}\left\{\frac{f(t) - f(t - \Delta T)}{\Delta T}\right\}$$

which is conventional integer order differentiation.

For general case, we take $q = 1/2$, then $\beta_{1/2} = \frac{\left(j - \frac{1}{2}\right)}{(j+1)}$, for all $j \geq 1 \to \beta_{1/2} > 0$.
It can be seen that β_q will be positive for all q when $j \geq 1$. It is also true that $0 > \alpha \geq -1$, for all q in the range. Therefore

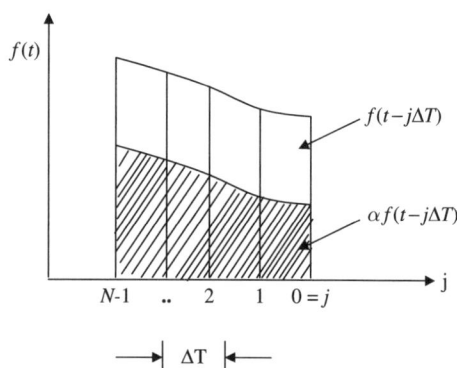

Fig. 5.9 Geometric interpretation if infinitesimal incremental fractional integration

5.8 Infinitesimal Element Geometrical Interpretation of Fractional Differintegrations

$$_aD_t^{1/2}f(t) = \lim_{\Delta T \to 0}\left\{\ldots \alpha \Delta T^{-1/2}\left[f(t - j\Delta T) + \beta_{1_{1/2}} f(t - (j+1)\Delta T)\right] + \ldots\right\}$$

For general q the expression is

$$_aD_t^q f(t) = \lim_{\Delta T \to 0}\left\{\ldots \alpha \Delta T^{-q}\left[f(t - j\Delta T) + \beta_q f(t - (j+1)\Delta T)\right] + \ldots\right\}$$

So after the first $j = 0$ term, it is seen that all terms are a direct sum of negatively weighted functions, again integration process. However, the effectiveness of the weighing ΔT^{-q} is changed as $q > 0$. The first $j = 0$ term for all q is $\Delta T^{-q} f(t)$, thus considering the first two terms, we have

$$_aD_t^q f(t) = \lim_{\Delta T \to 0}\left\{\frac{f(t) - qf(t - \Delta T)}{\Delta T^q} + \ldots\right\}$$

This brings in an effective differentiation (for $q > 0$) though scaled by ΔT^q instead of ΔT, as in the case of one-order differentiation.

If q is taken as 1, then this returns to the rate of change, like velocity (slope). For value 2, i.e., outside the range, it yields acceleration. Therefore for 0–1, the expression may be called as fractional rate of change of function. Figure 5.10 shows the $j = 0$ and $j = 1$ points of a geometric approximation to the qth derivative. The slope between the curves multiplied by ΔT^{-q+1} is loosely a geometric interpretation for this part of the fractional derivative or fractional rate. The remaining terms after the first two are like integration terms that may be interpreted as in Fig. 5.9, but meaningless at limit $\Delta T \to 0$.

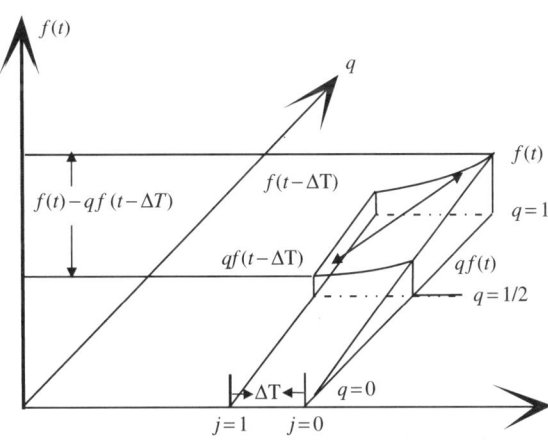

Fig. 5.10 Geometric interpretation of fractional differentiation for $j = 0$ to $j = 1$

5.9 Advance Digital Algorithms Realization for Fractional Controls

As seen from the definitions of fractional differintegrations, the operations are not a local or point property, rather distributed quantities. Therefore in the process to obtain value of fractional differentiation and integration, a bit of historical behavior is required. In spatial terms what is required is the behavior at the neighborhood of that function. The GL and RL definitions capture that historical behavior of the function, undoubtedly enormous amount of memory requirement is needed in spite of having short-memory principle. The historical behavior of the function is described by initialization function (instead of constant states in integer order calculus). Minimum 100 recent past history of the function is required to obtain good estimate with tolerable error and to get that initial function. There are advances in algorithms to emulate the fractional differintegration by digital control theory of discretization which leads to several other memory efficient methods, as compared to GL method described herein. Few advance digital control algorithms are power series expansion of Tustin rule, Al-Alouni rule, continued fraction expansion, and method of interpolation through fractional order delay. These advance algorithms are to realize the fractional control digital systems with approximately one-tenth of memory requirement as compared with the short-memory principle of the GL method. Advances in the algorithm development in this direction are ongoing process, and this century will see several of it. In this section, attempt is made to give the digital realization of the fractional differintegration operators; the readers, knowledge about z- transforms and control system basics and fundamentals of digital filters is required.

5.9.1 Concept of Generating Function

Discretization of fractional order differentiation s^r can be expressed by so-called generating function. Here s is the Laplace variable and discrete domain is z variable. To revise the concepts of digital controls, the monomial z^{-1} is the unit delay dictated by sampling or dicretization time T. The generating function is $s = \omega(z^{-1})$. For backward difference or Euler's rule, the function is $G_{gf}(z^{-1}) = \frac{1-z^{-1}}{T}$. Performing the power series expansion (PSE) of the $(1 - z^{-1})^{\pm r}$, using short-memory principle we get

$$(s)^{\pm r} = \left(G_{gf}(z^{-1})\right)^{\pm r} = T^{\mp r} z^{-[L/T]} \sum_{j=0}^{[L/T]} (-1)^j \binom{\pm r}{j} z^{[L/T]-j}$$

T is the sampling period, L is the Memory length, $(-1)^j \binom{\pm r}{j}$ are binomial coefficients of the form, i.e., $c_0^{(r)} = 1, c_j^{(r)} = \left(1 - \frac{1+(\pm r)}{j}\right) c_j^{(r)} - 1$. This PSE of

5.9 Advance Digital Algorithms Realization for Fractional Controls

the backward difference rule gives the digital realization of the GL method with short-memory principle. Applying this realization in digital filter realization gives the Finite Impulse Response (FIR) digital filter.

Generating function for the Trapezoidal rule (Tustin) is

$$G_T(z) = \left(G_{gf}(z^{-1})\right)^{\pm r} = \left(\frac{2}{T}\frac{1-z^{-1}}{1+z^{-1}}\right)^{\pm r}$$

Al-Alouni mixed Trapezoidal and Euler generating function is

$$G(z) = \left(G_{gf}(z^{-1})\right)^{\pm r} = \left(\frac{8}{7T}\frac{1-z^{-1}}{1+z^{-1}}\right)^{\pm r};$$

these are infinite order rational discrete time transfer functions. To approximate it with a finite order rational one, Continued Fraction Expansion (CFE) is an efficient way.

$$G(z) = \hat{G}(z) = a_0(z) + \cfrac{b_1(z)}{a_1(z) + \cfrac{b_2(z)}{a_2(z) + \cfrac{b_3(z)}{\ldots}}},$$

with a_i, b_i are either rational functions of the variable z or constants. With CFE method, the realization obtained is Infinite Impulse Response (IIR) digital filter. In number theory, the CFE method is used to represent a real number. A number can be represented as

$$x = a_0 + \cfrac{1}{a_1 + \cfrac{1}{a_2 + \ldots}},$$

for $x = \pi$, $a_0 = [\pi] = 3$, $a_1 = \left[\frac{1}{\pi-3}\right] = 7$, $a_2 = \left[\frac{1}{\frac{1}{\pi-3}-7}\right] = 15$.

The [*] is "FLOOR" function; it returns the integer part after operation.

5.9.2 Digital Filter Realization by Rational Function Approximation for Fractional Operator

Al-Alouni (1997) stated that magnitude of frequency of ideal integrator $1/s$ lies between that of Simpson and Trapezoidal digital integrators. It is reasonable to interpolate the Simpson and Trapezoidal digital integrators to compromise the high-frequency response. The Simpson digital integrator is

$$H_S(z) = \frac{T(z^2 + 4z + 1)}{3(z^2 - 1)},$$

the Tustin (trapezoidal) integrator is

$$H_T(z) = \frac{T}{2}\frac{(z+1)}{(z-1)}.$$

The combined digital integrator is $H(z) = aH_S(z) + (1-a)H_T(z)$, and the tuning knob fraction is $a \in [0, 1]$. Putting Simpson integrator and Tustin integrator, we obtain

$$H(z) = \frac{T(3-a)\{z^2 + [2(3+a)/(3-a)]z + 1\}}{6(z^2-1)} = \frac{T(3-a)(z+r_1)(z+r_2)}{6(z^2-1)}$$

with

$$r_1 = \frac{3+a+2\sqrt{3a}}{(3-a)}$$

and

$$r_2 = \frac{3+a-2\sqrt{3a}}{(3-a)}.$$

Note that $r_1 = 1/r_2$ and $r_1 = r_2 = 1$, when $a = 0$ (Pure Tustin). For $a \neq 0$, $H(z)$ must have one non-minimum phase zero.

First we can obtain a family of new integer order digital differentiators from the digital integrator introduced above by mixing Simpson and Tustin. Direct inversion of $H(z)$ will give an unstable filter since $H(z)$ has non-minimum phase (NMP) zero, r_1. By reflecting the NMP r_1 to $1/r_1$, i.e., r_2 we have approximate as

$$\hat{H}(z) = K\frac{T(3-a)(z+r_2)^2}{6(z^2-1)}.$$

To determine K, let the final value of the impulse response have $H(z)$ and $\hat{H}(z)$ be the same. Applying final value theorem, i.e., $\lim_{z \to 1}(z-1)H(z) = \lim_{z \to 1}(z-1)\hat{H}(z)$, gives $K = r_1$. Therefore, the new families of the digital differentiator are given by generating the function

$$G(z) = \frac{1}{\hat{H}(z)} = \frac{6(z^2-1)}{r_1T(3-a)(z+r_2)^2} = \frac{6r_2(z^2-1)}{T(3-a)(z+r_2)^2}$$

Finally, we can obtain the expression for the Digital Fractional Order Differentiator as

$$G(z^{-1}) = \left(G_{gf}(z^{-1})\right)^r = k_0\left(\frac{1-z^{-2}}{(1+bz^{-1})^2}\right)^r$$

Where $r \in [0, 1]$, $k_0 = \left(\frac{6r_2}{T(3-a)}\right)^r$ and $b = r_2$.

5.9 Advance Digital Algorithms Realization for Fractional Controls

It is well known that compared to Power Series Expansion (PSE) the Continued Fraction Expansion (CFE) is a method of evaluation of functions with faster convergence in larger domain in complex plane. Using the CFE, an approximation for an irrational function $G(z^{-1})$ can be expressed as approximation. The CFE can be automated by using MATLAB symbolic toolbox

$$CFE\left(\frac{1-x^2}{(1+bx)^2}\right)^r$$

with $x = z^{-1}$ for desired order n. The MATLAB script generates the above CFE with polynomial containing the numerator and denominator with coefficients being function of b and r. Following list is IIR transfer function for $n = 3$, $a = 0 - 1$ in steps of 0.25 for semi-differentiation $r = 0.5$ at sampling time of $T = 0.001s$ (1 ms).

$G_{n,a}$ means nth order polynomial approximate at a mixing value a.

$$G_{(3,0)}(z^{-1}) = \frac{357.8 - 178.9z^{-1} - 178.9z^{-2} + 44.72}{8 + 4z^{-1} - 4z^{-2} - z^{-3}}$$

$$G_{(3,0.25)}(z^{-1}) = \frac{392.9 - 78.04z^{-1}349.8z^{-2} + 88.97z^{-3}}{11.32 + 4z^{-1} - 5.66z^{-2} - z^{-3}}$$

$$G_{(3,0.5)}(z^{-1}) = \frac{1501 - 503.6z^{-1} - 1289z^{-2} + 446.5z^{-3}}{47.26 + 4z^{-1} - 23.63z^{-2} - z^{-3}}$$

$$G_{(3,0.75)}(z^{-1}) = \frac{968.1 - 442z^{-1} - 820.8z^{-2}363z^{-3}}{32.47 - 4z^{-1} - 16.24z^{-2} + z^{-3}}$$

$$G_{(3,1.00)}(z^{-1}) = \frac{353.1 - 208z^{-1} - 297.4z^{-2} + 164.7z^{-3}}{12.46 - 4z^{-1} - 6.228z^{-2} + z^{-3}}$$

The fourth-order approximation is for digital IIR is listed below

$$G_{(4,0)}(z^{-1}) = \frac{715.5 - 357.8z^{-1} - 536.7z^{-2} + 178.9z^{-3} + 44.72z^{-4}}{16 + 8z^{-1} - 12z^{-2} - 4z^{-3} + z^{-4}}$$

$$G_{(4,0.25)}(z^{-1}) = \frac{555.3 - 392.9z^{-1} - 477.2z^{-2} + 349.8z^{-3} - 19.56z^{-4}}{16 - 2.489z^{-1} - 12z^{-2} + 1.245z^{-3} + z^{-4}}$$

$$G_{(4,0.5)}(z^{-1}) = \frac{508.1 - 1501z^{-1} - 4.478z^{-2} + 1289z^{-3} - 382.9z^{-4}}{16 - 40.54z^{-1} - 12z^{-2} + 20.27z^{-3} + z^{-4}}$$

$$G_{(4,0.75)}(z^{-1}) = \frac{477 + 968.1z^{-1} - 919z^{-2} - 820.8z^{-3} + 422.7z^{-4}}{16 + 37.8z^{-1} - 12z^{-2} - 8.371z^{-3} + z^{-4}}$$

5.9.3 Filter Stability Consideration

For odd CFE $n = 3$, the pole-zero maps are nicely placed and behaved, that is, all pole and zeros of the transfer function lie inside the unit circle and the poles and zeros are interlaced along the segment of the real axis corresponding to $z \in (-1, 1)$. However, when $n = 4$ and even if "a" is near unity (tending toward Simpson rule), there may be one canceling pole-zero pair, which may not be desirable. Therefore suggestion is to use odd order (polynomial) expansions for CFE. When $a = 0$ for Tustin (trapezoidal) rule pole-zero (P-Z) maps always inside unit circle in an interlacing way along the segment of the real axis $z \in (-1, 1)$. For special case $a = 0$, Tustin CFE for odd polynomial expansion gives

$$D^r(z) = CFE\left(\frac{1-z^{-1}}{1+z^{-1}}\right)^r = 1 + \cfrac{z^{-1}}{-\frac{1}{2}\frac{1}{r} + \cfrac{z^{-1}}{-2 + \cfrac{z^{-1}}{\frac{3}{2}\frac{r}{r^2-1} + \frac{z^{-1}}{2 + \ldots}}}}$$

For semi-differentiation realization $r = 0.5$ for discretization time $T = 0.001s$ (1 ms), the approximate odd CFE expansions are

$$G_1(z) = 44.72\frac{z - 0.5}{z + 0.5}, \quad G_3(z) = 44.72\frac{z^3 - 0.5z^2 - 0.5z + 0.125}{z^3 + 0.5z^2 - 0.5z - 0.125}$$

5.10 Local Fractional Derivatives

Fractals and multifractals functions and corresponding curves or surfaces are found in numerous places in non-linear and non-equilibrium phenomenon, for example, low viscous turbulent fluid, Brownian motion. These phenomena give occurrence of continuous but highly irregular (non-differentiable) curves. However the precise nature of connection between the dimensions of the graph of fractal curve and fractional differentiability property was only recognized recently. K.M. Kolwankar and Anil D. Gangal introduced a new notion of Local Fractional Derivative (LFD).

In this section, only the definition of this LFD will be introduced. The classical definition of fractional derivatives of Reimann–Liouvelli (RL), Grunwald–Letnikov (GL) discussed in detail in this book makes it a non-local property. Also as discussed the RL approach of fractional derivative of a constant makes it non- zero. These two features make extraction of scaling information somewhat difficult. LFD tries to overcome these issues by having a neighborhood point approach as defined:

$$D^q f(y) = \lim_{x \to y} \frac{d^q (f(x) - f(y))}{d(x - y)^q}$$

If this limit exists and is finite, then we say the LFD of order $q(0 < q < 1)$ at $x = y$ exists. In this definition, the lower limit y is treated as a constant. The subtraction of $f(y)$ corrects for the fact that fractional derivative of a constant is not zero. Whereas, limit $x \to y$ is taken to remove non-local contents. Advantage of defining LFD in this manner lies in its local nature and hence allowing the study of pointwise behavior of functions.

5.11 Concluding Comments

The physical and geometric interpretation of fractional differintegration process has shown some insight of the mathematics, which lies in between complete integration and complete differentiation. The elaborate block diagrams provide understanding for computation of these fractional processes. The upcoming field of local fractional derivatives is just introduced, which is a tool for description of fractal process. The concepts of minimizing computation effort by digital signal processing fundamentals have given a direction to evolve efficient algorithms for digital control science applications. The concepts are evolving even today, and future will see much more insight into the concepts of fractional differintegrations.

Chapter 6
Initialized Differintegrals and Generalized Calculus

6.1 Introduction

This chapter demonstrates the need for a non-constant initialization for the fractional calculus. Here basic definitions are formed for the initialized fractional differintegrals (differentials and integrals). Here two basic popular definitions of fractional calculus are considered: Riemann–Liouville (RL) and Grunwald–Letnikov (GL). Two forms of initialization methods are prevalent, the "terminal initialization" and the "side initialization." The issue of initialization has been an essentially neglected subject in the development of the fractional calculus. Liouville's choice of lower limit as $-\infty$ and Riemann's choice as c were in fact issues related to the same initialization. Ross and Caputo maintained that to satisfy the composition of the fractional differintegrals, the integrated function and its integer order derivatives must be zero, for all times up to and including the start of fractional differintegration. Ross stated that "The greatest difficulty in Riemann's theory is the interpretation of complimentary function. The question of existence of complimentary function caused much of confusion. Liouvelli was led to error and Riemann became inextricably entangled in his concept of a complimentary function." The complimentary function issue is raised here because an initialization function, "which accounts for effect of history," of the function, for fractional derivatives and integrals, will appear in the definitions of this chapter. The form of initialization function is kept similar to what Riemann has used as complimentary function $\psi(x)$; however, its meaning and use is different.

Constant initialization of the past is insufficiently general, the widely used contemporary equations for the Laplace transform for differintegrals based on that assumption also lacks generality. Therefore, the generalized form is presented here. In solution of fractional differential equations with assumed history, the set of initializing constants representing the values of fractional differintegrals at $t = 0$, that are ineffective, will be deliberated in this chapter. Therefore, "non-constant initialization" for generalized concept of integration and differentiation is required.

Also the fundamental fractional order differential equation concept is touched, its solution is the fundamental time response, whose combination provide solution to complicated systems. From this transfer function is constructed the fractional pole, which is the transfer function of the fundamental fractional differential equation, and is the fundamental building block for more complicated fractional order systems.

A brief discussion on criteria and properties of generalized calculus as given by Ross (1974) is given and then simple examples are provided for getting the gist of fractional calculus, with importance given to initialization.

6.2 Notations of Differintegrals

Mathematicians have used several notations since the birth of fractional calculus. Several contemporary notations for fractional differentiation and fractional integration are mentioned in Sect. 1.4. Here, attempt will be made to standardize the notations as differintegrals. The same operator is used as an integrator when the index is negative and as a differentiator when the index is positive. Separate notations will be used to indicate initialized differintegral operator and un-initialized operator.

$_cD_t^q f(t)$ represents "initialized" qth order differintegration of $f(t)$ from start point c to t $_cd_t^q f(t)$ represents "un-initialized" generalized (or fractional) qth order differintegral. This is also same as

$$\frac{d^q f(t)}{[d(t-c)]^q} \equiv {_cd_t^q} f(t),$$

shifting the origin of function at start of the point from where differintegration starts. This un-initialized operator can also be short formed as $d^q f(t)$. The initialization function (not a constant) is represented as $\psi(f, q, a, c, t)$, meaning that this is function of the independent variable t, and is for differintegral operator of order q, the function born at $t = 0$ (before that, the function is zero), and differintegral process starting at c. This initialization function can be short formed as $\psi(t), \psi(f, q, t)$. Therefore, the expression between initialized differintegral and un-initialized one is

$$_cD_t^q f(t) = {_cd_t^q} f(t) + \psi(f, q, a, c, t)$$

The notation contains lower limit of the process at the front subscript and the order of the process at the tail superscript, with independent variable with respect to what is being differintegrated.

6.3 Requirement of Initialization

In real applications, it is usually the case that the problem to be solved is in some way isolated from the past. That is, it should not be necessary to retreat to $-\infty$ in time to start the analysis. Usually, the analyst desires to start the analysis at some time $t = t_0$, with the knowledge (or assumption) of all values of the function and its derivatives, specifically $f(t_0), f^{(1)}(t_0), f^{(2)}(t_0)\ldots\ldots f^{(n)}(t_0)$, in the case of integer order calculus. In modern parlance, this collection is called "state" and contains the

6.3 Requirement of Initialization

effect of all the past history. One way in which the behavior of the semi-infinite transmission line can be described is in terms of its input behavior (impedance) at the open end of the line, that is, as semi-differential equation. However, to practically use such fractional order differential equation requires additional function of time. In terms of the physics, this time function relates back to the initial voltage distribution (distributed initialization) on the semi-infinite lossy line. From Chap. 3, the input terminal behavior of the same is

$$V(0,s) = \frac{rI(0,s)}{\sqrt{\frac{s}{\alpha}}} + \frac{1}{\alpha\sqrt{\frac{s}{\alpha}}}\int_0^\infty e^{\sqrt{\frac{s}{\alpha}}\lambda} V(\lambda,0)d\lambda,$$

where λ is the dummy variable of integration and s is Laplace variable. The above expression, in Chap. 3, was obtained by iterated Laplace transformation technique applied to the basic diffusion equation. However, an attraction of the fractional calculus is the ability to express the behavior of the line (a distributed system or mathematically partial differential equation) as part of the system of distributed equations using fractional differential equations. Such a fractional differential equation for the semi-infinite lossy line is

$$\frac{d^{1/2}v(t)}{dt^{1/2}} = r\sqrt{\alpha}i(t),$$

assuming $v(x,0) = 0$. To initialize this distributed system, a function of time $\psi(t)$ must be added to account for the integral term of the obtained expression for $V(0,s)$, written above. With, this the fractional differential equation is

$$\frac{d^{1/2}v(t)}{dt^{1/2}} + \psi(t) = r\sqrt{\alpha}i(t).$$

The focus in this chapter will be on $\psi(t)$. Clearly, one can addend such terms in ad hoc way to the fractional differential equations which are being solved; the formal approach to evaluate this function is presented in detail in this chapter. If the analyst is constrained that the initial function value and all its derivatives are zero, the range of applicability for this entire class of problems, which includes eventually all distributed systems, will greatly be limited. Therefore, all fractional ordinary differintegral equations require initialization terms to be associated with each fractional differintegration term, in order to complete the description. This requirement is a generalization to the requirement of a set of initialization constants "states," in integer order ordinary differential equations. Fundamentally, it is the information to start the integration process of the differential equations while properly accounting for the effect of the past.

6.4 Initialization Fractional Integration (Riemann–Liouville Approach)

This non-constant; initialization function $\psi(t)$ which shall be elucidated clearly brings out the past history, also brings to the definition of the fractional integral the effect of past, namely, effect of fractionally integrating the function from its birth. This added effect will also be influencing the process after time t, the start of integration process.

Consider fractional order q integration of the $f(t)$, the first starting at $t = a$ and second starting at $t = c > a$

$$_a d_t^{-q} f(t) = \frac{1}{\Gamma(q)} \int_a^t (t-\tau)^{q-1} f(\tau) d\tau \tag{6.1}$$

$$_c d_t^{-q} f(t) = \frac{1}{\Gamma(q)} \int_c^t (t-\tau)^{q-1} f(\tau) d\tau \tag{6.2}$$

Assume that the function was born at $t = a$, that is, $f(t) = 0$ for all time less than equal to a, i.e., $f(t) = 0, t \le a$. Then the time period between a and c may be considered as history. The assumption is that the integral $\left(_c d_t^{-q} f(t)\right)$ is properly initialized so that it should function as continuation of integral starting at $t = a$. To this, therefore, an initialization must be added (to $_c d_t^{-q} f(t)$) so that the fractional integral starting at $t = c$ should be identical to the result starting at $t = a$ for $t > c$. We call what Riemann proposed as complimentary function as initialization function as ψ.

We have for the above argument the following:
$_c d_t^{-q} f(t) + \psi = {_a d_t^{-q}} f(t), t > c$. Then, $\psi = {_a d_t^{-q}} f(t) - {_c d_t^{-q}} f(t), t > c$
Therefore,

$$\psi = \frac{1}{\Gamma(q)} \int_a^c (t-\tau)^{q-1} f(\tau) d\tau \equiv {_a d_c^{-q}} f(t). \quad t > c \tag{6.3}$$

Here ψ is independent of t, thus is "non-constant." For integer order integration, we put $q = 1$, and see that $\psi = \int_a^c f(\tau) d\tau = K$, a constant. Because of increased complexity of the initialization relative to the integer order calculus, it is important to formalize the initialization process. This formalization will include the initialization term into the definition of these fundamental fractional order calculus operators.

Two types of initialization are considered! (1) the terminal initialization, where it is assumed that the differintegral operator can be initialized (charged) by effectively differintegrating prior to the start time, $t = c$ and (2) the side initialization, where fully arbitrary initialization may be applied to the differintegral operator at

6.4 Initialization Fractional Integration (Riemann–Liouville Approach)

time $t = c$. These are in contemporary terms and may be stated as terminal charging and side charging. First, we restrict to RL type of differintegrals for formalizing these definitions. This initialization function ψ has the effect of allowing the function $f(t)$ and its derivatives to start at a value other than zero, namely ${}_aD_c^{-q}f(t)_{at\ t=c}$, and continues to contribute to differintegral response after $t = c$. That is, a function of time is added to the uninitialized integral (not just a constant at $t = c$).

6.4.1 Terminal Initialization

The standard contemporary definition of fractional integral (RL) is accepted only if the differintegrand $f(t) = 0$ for all $t \leq a$. The initialization period (or space) is region $a \leq t \leq c$. The fractional integration takes place for $t > c \geq a$. Furthermore, the fractional integration starts at $t = c$ (i.e. point of initialization).

$$_aD_t^{-q}f(t) \equiv \frac{1}{\Gamma(q)}\int_a^t (t-\tau)^{q-1}f(\tau)d\tau, \quad q \geq 0. \text{ and } t > a \qquad (6.4)$$

subject to $f(t) = 0$ for all $t \leq a$.

The following definition of fractional integration will apply generally (at any $t > c$):

$$_cD_t^{-q}f(t) \equiv \frac{1}{\Gamma(q)}\int_c^t (t-\tau)^{q-1}f(\tau)d\tau + \psi(f, -q, a, c, t), \qquad (6.5)$$

$q \geq 0, t > a, c \geq a$, and $f(t) = 0$ at $t \leq a$.

The function $\psi(f, -q, a, c, t)$ is called the initialization function and will be chosen such that

$$_aD_t^{-q}f(t) = {}_cD_t^{-q}f(t). \quad t > c \qquad (6.6)$$

This condition gives the following:

$$\frac{1}{\Gamma(q)}\int_a^t (t-\tau)^{q-1}f(\tau)d\tau = \frac{1}{\Gamma(q)}\int_c^t (t-\tau)^{q-1}f(\tau)d\tau + \psi(f, -q, a, c, t). \qquad (6.7)$$

Since, $\int_a^t g(\tau)d\tau = \int_a^c g(\tau)d\tau + \int_c^t g(\tau)d\tau$.

Therefore, we get

$$\psi(f, -q, a, c, t) = {}_aD_c^{-q}f(t) = \frac{1}{\Gamma(q)}\int_a^c (t-\tau)^{q-1}f(\tau)d\tau \qquad (6.8)$$

$t > c$ and $q > 0$

This expression for $\psi(t)$ gives "terminal initialization," and also brings out in the definition of fractional integral the effect of the past "history," namely the effect of fractionally integrating the $f(t)$ from a to c. This effect is also called terminal charging.

6.4.2 Side Initialization

When ψ is arbitrary and terminal initialization equation is not valid, then the effect is called "side initialization," or side charging. Figure 6.1 demonstrates the concept initialization as a block diagram, as a signal flow graph.

Example 1. Let function $f(t) = t$ for $t > 0$ and $f(t) = 0$ for $t < 0$. The semi-integral process with initialization is demonstrated below from the start point of integration at $t = 1$. By applying RL formulations for fractional integration, we obtain the following:

$$_0D_t^{-1/2}t = \frac{1}{\Gamma(1/2)} \int_0^t (t-\tau)^{0.5-1}\tau.d\tau = \frac{1}{\Gamma(1/2)} \int_t^0 \frac{(t-x)(-dx)}{x^{1/2}}$$

$$= \frac{t^{3/2}}{\frac{3}{2}\cdot\frac{1}{2}\Gamma(1/2)} = \frac{4}{3\sqrt{\pi}}t^{3/2}$$

$$_1D_t^{-1/2}t = \frac{1}{\Gamma(1/2)} \int_{t-1}^0 \frac{(t-x)(-dx)}{x^{1/2}} + \psi(t, -0.5, 0, 1, t)$$

$$= \frac{2}{3\sqrt{\pi}}\left[(t-1)^{1/2}(2t+1)\right] + \psi(t, -1/2, 0, 1, t)$$

$$t > 1, \&, \Gamma(0.5) = \sqrt{\pi}$$

$$\psi(t, -1/2, 0, 1, t) = \frac{2}{3\sqrt{\pi}}\left[2t^{3/2} - (2t+1)(t-1)^{1/2}\right]$$

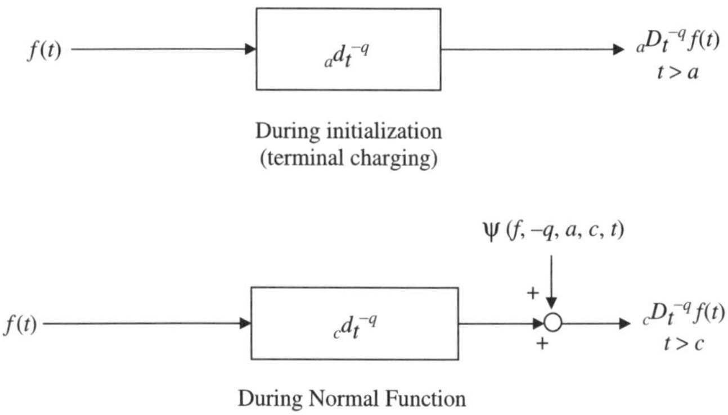

$f(t)$ ⟶ [$_ad_t^{-q}$] ⟶ $_aD_t^{-q}f(t)$, $t > a$

During initialization
(terminal charging)

$\psi(f, -q, a, c, t)$

$f(t)$ ⟶ [$_cd_t^{-q}$] ⟶ + ⊕ + ⟶ $_cD_t^{-q}f(t)$, $t > c$

During Normal Function

Fig. 6.1 Signal flow graph for demonstrating initialization of fractional integration

6.5 Initializing Fractional Derivative (Riemann–Liouvelle Approach)

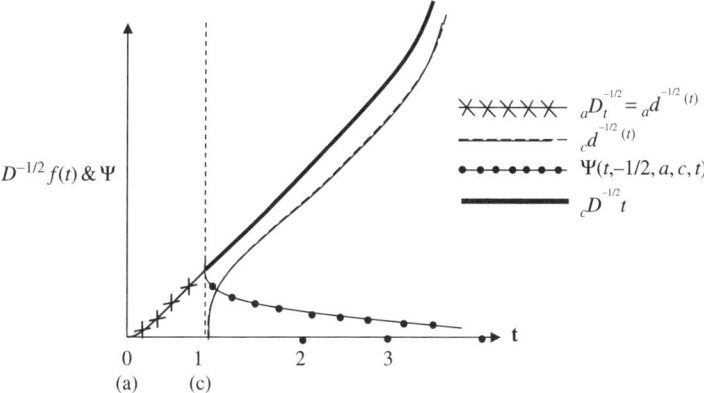

Fig. 6.2 Graphical representation of initialized differintegration for $f(t) = t$

The nature of initialization function may be noted, which is semi-integration of the function from 0 to 1, and is function of t. Also as time passes, the function decays to zero, i.e. "the effect of history is forgotten as future grows!" $\lim_{t \to \infty} \psi(t, -0.5, 0, 1, t) = 0$, at least in this particular example. This is shown in the Fig. 6.2.

6.5 Initializing Fractional Derivative (Riemann–Liouvelle Approach)

This is contrary to integer order derivative, which is a point and also a local quantity (property), where initialization is not called for. However, the definitions of fractional derivatives do contain fractional integration, and thus fractional derivative of a function is not a point quantity. For fractional derivative the initialization process is called for. Fractional derivative is a non-local operator and therefore has history. In solution of differential equations, the initialization constants, which sets the initial values of the derivatives, really have the effect of accounting for the integration of the derivative from $-\infty$ to starting time of the integration (of the differential equation). In the fractional calculus, initialization for derivatives are also required for handling the effect of "distributed initialization," in actual system. Distributed initialization means initial voltage/charge profile in semi-infinite distributed transmission line (lossy of order $1/2$ or lossless of order zero), or in case of initial strain distribution in elastic semi-infinite bar (order one). Extending generalization concept, the integer order derivative also calls for initialization in "fractional context." Thus, a generalized integer order differentiation is defined as (with initialization) $t > c$.

$$_cD_t^m f(t) \equiv \frac{d^m}{dt^m} f(t) + \psi(f, m, a, c, t). \tag{6.9}$$

Here m is positive integer and $\psi(f, m, a, c, t)$ is an initialization function. Now bare or un-initialized fractional derivative is defined as:

$$_aD_t^q f(t) \equiv {_aD_t^m} {_aD_t^{-p}} f(t). \tag{6.10}$$

$q \geq 0$, $t > a$, and $f(t) = 0$ at $t \leq 0$, $q = m - p$. Meaning, m is the integer just greater than the fractional order q, by amount p. The function is born at $t = a$, and before that, the value is zero. The differentiation starts at $t > c$. Now as in fractional integration case, $\psi(f, -p, a, a, t) = 0$. Further more, consider $h(t) = {_aD_t^{-p}} f(t)$, i.e., fractional integral of function starting at a with initialized term $\psi(h, m, a, a, t) = 0$. The initialized fractional derivative is defined for $q \geq 0$ and $t > c \geq a$:

$$_cD_t^q f(t) \equiv {_cD_t^m} {_cD_t^{-p}} f(t). \tag{6.11}$$

6.5.1 Terminal Initialization

The definition and concept is similar to that obtained as terminal initialization for fractional integrals. The requirement is also the same as for the fractional integrals, that is

$$_cD_t^q f(t) = {_aD_t^q} f(t) \tag{6.12}$$

for all $t > c \geq a$. Specifically, this requires compatibility of the derivatives starting at $t = a$ and $t = c$ for all $t > c$. Therefore, it follows that

$$_cD_t^m {_cD_t^{-p}} f(t) = {_aD_t^m} {_aD_t^{-p}} f(t) \tag{6.13}$$

Expanding the fractional integrals with initialization we obtain

$$_cD_t^m \left(\frac{1}{\Gamma(p)} \int_c^t (t-\tau)^{p-1} f(\tau) d\tau + \psi(f, -p, a, c, t) \right)$$

$$= {_aD_t^m} \left(\frac{1}{\Gamma(p)} \int_a^t (t-\tau)^{p-1} f(\tau) d\tau + \psi(f, -p, a, a, t) \right) \tag{6.14}$$

for $t > c$ and $\psi(f, -p, a, a, t) = 0$. Using the definition of generalized integer order derivative as defined above we get

$$\frac{d^m}{dt^m} \left\{ \frac{1}{\Gamma(p)} \int_c^t (t-\tau)^{p-1} f(\tau) d\tau + \psi(f, -p, a, c, t) \right\} + \psi(h_1, m, a, c, t)$$

$$= \frac{d^m}{dt^m} \frac{1}{\Gamma(p)} \int_a^t (t-\tau)^{p-1} f(\tau) d\tau + \psi(h_2, m, a, a, t) \tag{6.15}$$

6.5 Initializing Fractional Derivative(Riemann–Liouvelle Approach)

where $h_1 = {_c}D_t^{-p}f(t)$ and $h_2 = {_a}D_t^{-p}f(t)$; the integer order derivative is initialized at $t = a$, thus $\psi(h_2, m, a, a, t) = 0$. After rearranging the integrals we get

$$\psi(h_1, m, a, c, t) = \frac{d^m}{dt^m}\left(\frac{1}{\Gamma(p)}\int_a^c (t-\tau)^{p-1}f(\tau)d\tau - \psi(f, -p, a, c, t)\right). \tag{6.16}$$

This is, the expression and "the requirement for the initialization for the derivative in general."

Under the condition of terminal charging of the fractional integral

$$\psi(f, -p, a, c, t) = \frac{1}{\Gamma(p)}\int_a^c (t-\tau)^{p-1}f(\tau)d\tau,$$

and is the initialization function of fractional integration as defined and derived earlier (6.8). Therefore, $\psi(h_1, m, a, c, t) = 0$, a very important result is seen, that is, "integer order differentiation cannot be initialized through the terminal (terminal charging)."

6.5.2 Side Initialization

Refer the expression (6.16) and the requirement for initialization for the general derivative, as obtained in the terminal charging case. If side charging is employed, then the function $\psi(f, -p, a, c, t)$ is arbitrary. Thus it can be inferred from the requirement equation (6.16) that $\psi(f, -p, a, c, t)$ or $\psi(h_1, m, a, c, t)$ can be arbitrary, but not together, but should also satisfy the requirement expression derived above.

The generalized expression for the side charging case can be stated as:

$$_cD_t^q f(t) = {_c}D_t^m\left\{\frac{1}{\Gamma(p)}\int_c^t (t-\tau)^{p-1}f(\tau)d\tau + \psi(f, -p, a, c, t)\right\}, (t. > c) \tag{6.17}$$

m is positive integer $> q$, with $q = m - p$

$$_cD_t^q f(t) = \frac{d^m}{dt^m}\frac{1}{\Gamma(p)}\int_c^t (t-\tau)^{p-1}f(\tau)d\tau + \frac{d^m}{dt^m}\psi(f, -p, a, c, t)$$
$$+ \psi(h, m, a, c, t) \tag{6.18}$$

where $h(t) = {_a}D_t^{-p}f(t)$. Here, both the initialization terms are arbitrary and thus may be considered as a single (arbitrary) term, namely

$$\psi(f, q, a, c, t) \equiv \frac{d^m}{dt^m}\psi(f, -p, a, c, t) + \psi(h, m, a, c, t). \tag{6.19}$$

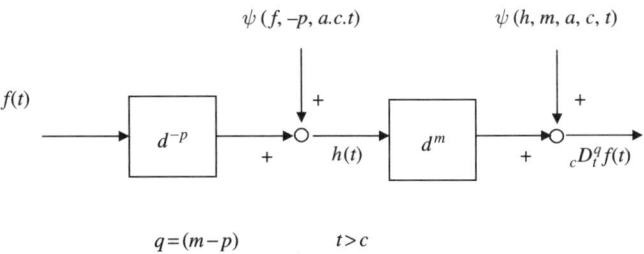

Fig. 6.3 Initialization of fractional derivative

In case of terminal charging, the fractional integral initialization part

$$\psi(f, -p, a, c, t) = \frac{1}{\Gamma(p)} \int_a^c (t - \tau)^{p-1} f(\tau) d\tau \text{ for } t > c \quad (6.20)$$

Figure 6.3 demonstrates the initialization concept for fractional derivative

6.6 Initializing Fractional Differintegrals (Grunwald–Letnikov Approach)

Here in this approach too, take the function's starting time as a, and the differinteration process starts at $t = c$. An initialization (notation same as for RL approach) is introduced to account for past history and goes back to $t = a$, with $f(t) = 0$ at all time before $t = a$. Then differintegration with arbitrary order q is

$$_aD_t^q f(t) = \frac{d^q f(t)}{[d(t-a)]^q} \equiv \lim_{N \to \infty} \frac{\left(\frac{t-a}{N}\right)^{-q}}{\Gamma(-q)} \sum_{j=0}^{N-1} \frac{\Gamma(j-q)}{\Gamma(j+1)} f\left(t - j\left\{\frac{t-a}{N}\right\}\right) \quad (6.21)$$

$t > a$ and $f(t) = 0$ at $t \le a$.

Grunwald–Letnikov (GL) definition for differintegrals will generally apply as:

$$_cD_t^q f(t) \equiv \frac{d^q f(t)}{[d(t-c)]^q} + \psi(f, q, a, c, t). \quad (6.22)$$

$f(t) = 0$ (at $t \le a$) and $c \ge a$. Here again, $\psi(f, q, a, c, t)$ is selected such that $_cD_t^q f(t)$ will produce the same result as $_aD_t^q f(t)$ for $t > c$. Expressing as

$$_cD_t^q f(t) = _cd_t^q f(t) + \psi(f, q, a, c, t) = _aD_t^q f(t) \quad (6.23)$$

will be self-explanatory, for all $t > c$, and $f(t) = 0$ for all $t \le a$. Therefore, $\psi(f, q, a, c, t) = _aD_t^q f(t) - _cd_t^q f(t)$ or identifying that $_aD_t^q f(t) \to _ad_t^q f(t)$, i.e., un-initialized differintegral, as per standard notation, we write the same as: $\psi(f,q,a,c,t) = {_a}d_t^q f(t) - {_c}d_t^q f(t)$. In this, substituting the GL series, we obtain the following:

$$\psi(f,q,a,c,t) = \lim_{N_1 \to \infty} \left\{ \frac{\left(\frac{t-a}{N_1}\right)^{-q}}{\Gamma(-q)} \sum_{j=0}^{N_1-1} \frac{\Gamma(j-q)}{\Gamma(j+1)} f\left(t - j\frac{t-a}{N_1}\right) \right\}$$

$$- \lim_{N_2 \to \infty} \left\{ \frac{\left(\frac{t-c}{N_2}\right)^{-q}}{\Gamma(-q)} \sum_{j=0}^{N_2-1} \frac{\Gamma(j-q)}{\Gamma(j+1)} f\left(t - j\frac{t-c}{N_2}\right) \right\} \quad (6.24)$$

For all $t > c$ and $f(t) = 0$ for $t < a$.

After considerable manipulations, by adjusting delay element as equal, that is $N_2 = ((t-c)/(t-a))N_1$, and adjusting with $\Delta T = (t-a)/N_1$ and $N_3 = ((c-a)/(t-a))N_1$

$$\psi(f,q,a,c,t) = \lim_{N_1 \to \infty} \left\{ \frac{\Delta T^{-q}}{\Gamma(-q)} \sum_{j=0}^{N_3-1} \frac{\Gamma(N_1-1-q-j)}{\Gamma(N_1-j)} f\left(t - [N_1-1-j]\Delta T\right) \right\}$$

(6.25)

6.7 Properties and Criteria for Generalized Differintegrals

One of the fundamental problems of fractional calculus is the requirement that the function and its derivatives be identically equal to zero at the start of initialization (i.e. start of differintegration process) at time $t = c$. This needed to assure composition or index law holds implying that ${_c}D_t^v {_c}D_t^u f(t) = {_c}D_t^u {_c}D_t^v f(t) = {_c}D_t^{u+v} f(t)$. It is difficult in engineering sciences to always require that the functions and its derivatives be at zero (rest) at initialization instants. This fundamentally implies that "there can be no initialization or composition is lost". Thus, it is not in general true that $f - \frac{d^{-Q}}{dt^{-Q}} \frac{d^Q f}{dt^Q} = 0$.

Therefore, while solving a fractional differential equation of the form $\frac{d^Q f}{dt^Q} = F$, additional terms must be added, like

$$f - \frac{d^{-Q}}{dt^{-Q}} \frac{d^Q f}{dt^Q} = C_1 t^{Q-1} + C_2 t^{Q-2} + \ldots C_m t^{Q-m},$$

to achieve the most general solution:

$$f = \frac{d^{-Q} F}{dt^{-Q}} + C_1 t^{Q-1} + C_2 t^{Q-2} + \ldots + C_m t^{Q-m}.$$

These issues described says that the index law or the composition law is inadequate.

Minimal set criteria have been thought fit to be applied for fractional (or generalized) calculus. They are listed as follows and are called Ross (1974) criteria:

i. If $f(z)$ is the analytic function of the complex variable z, the differintegral $_cD_z^v f(z)$ is the analytic function of z and v.
ii. The operator $_cD_x^v f(x)$ must produce the same result of differentiation, when v is a positive integer.
iii. If v is a negative integer (say $v = -n$), then $_cD_x^{-n} f(x)$ must produce the same result of n-fold integration of function $f(x)$, and $_cD_x^{-n} f(x)$ must vanish along with $f^{(1)}, f^{(2)}, \ldots, f^{(n-1)}$, all the $(n-1)$ derivatives at $x = c$.
iv. "Zero" operation leaves the function unchanged.

$$_cD_x^0 f(x) = f(x)$$

v. Linearity of the fractional (generalized) differintegral operator:

$$_cD_x^{-q} [af(x) + bg(x)] = a\,_cD_x^{-q} f(x) + b\,_cD_x^{-q} g(x)$$

vi. The law of exponents for arbitrary order holds

$$_cD_x^{-u}\,_cD_x^{-v} f(x) = {_cD_x^{-u-v}} f(x) = {_cD_x^{-v-u}} f(x)$$

The above notations are used by Ross.

It should be noted that there is a minor conflict contained in these criteria. Also a clear notation explanation should be given as $_cD_x^q f(x)$ in the above criteria is an un-initialized differintegral. It is correct as the function itself starts at c, and before that, the same is zero. So at $t = c$, $_cD_x^q f(x) = {_cd_x^q} f(x)$. The criterias (ii) and (iii) call for backward compatibility and the criteria (vi) calls for index law to be holding vis-à-vis integer order calculus.

The fundamental theorem of integer order calculus violates this "zero law" as: $d^{-m} d^m f(x) \neq d^0 f(x) = f(x)$, for all $f(x)$ and for all m (integer). The fundamental theorem states that $\int_c^t f'(t) = f(t) - f(c)$, and can be thus observed that the reversal of differentiation and integration differs from $f(t)$ by $f(c)$, that is by initialization (constant in integer order calculus). This failure in backward compatibility and index law is handled in the integer order calculus by constant of integration and by complimentary function for solution of differential equations (in ad hoc manner). The law of exponents (index law) is demonstrated in Fig. 6.4.

The discussion in all the differintegrations is limited to the real domain. Under the condition of terminal charging of the uth and vth differintegrations,

$$_cD_t^u\,_cD_t^v f(t) = {_cD_t^v}\,_cD_t^u f(t) = {_cD_t^{u+v}} f(t) \text{ for } t > 0$$

6.7 Properties and Criteria for Generalized Differintegrals

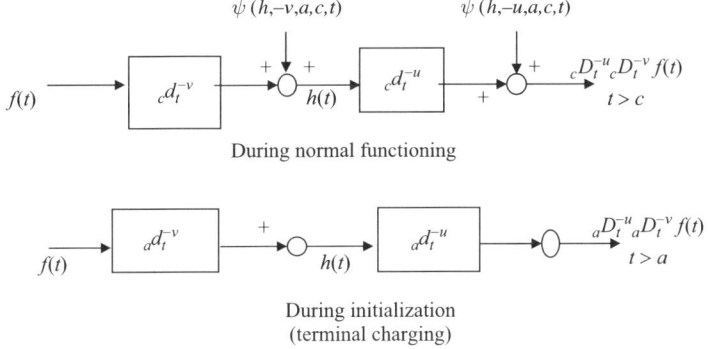

Fig. 6.4 Demonstration of index law

under the following conditions:

a. $u < 0, v < 0$ for continuous $f(t)$
b. $u > 0, v > 0$. For $f(t)$ is m times differentiable. $_aD_t^m f(t)$ exists and is non-zero continuous function of t for $t > a$, where m is an integer larger than integer part $[u]$ or $[v]$.
c. $u < 0, v > 0$ same as (b)

6.7.1 Terminal Charging

Under the conditions of terminal charging, the above properties and criterias holds; this provides credibility to the initialized fractional (generalized) calculus. Some conditions are however imposed, say on the linearity of fractional integrals. $_cD_t^{-v}(bf(t) + kg(t)) = b_cD_t^{-v}f(t) + k_cD_t^{-v}g(t), (t > c)$ holds only if $\psi(bf + kg, -v, a, c, t) = b\psi(f, -v, a, c, t) + k\psi(g, -v, a, c, t)$.

Relative to the criteria of backward compatibility with the integer order calculus, the addition of the initialized function is clearly a generalization relative to integer order calculus. In a strict sense $\psi(t) \neq 0$ violates the criteria (ii); however, we are looking for generalization of integer order calculus, and it is clear that this generalization (i.e. addition of initialization function) will be very useful in many applications.

Relative to the criteria of zero-order property holds for terminal charging.
Relative to linearity holds for the terminal charging subject to the above said rule. Relative to composition rule, the above (a) (b) and (c) should follow.

It is noted that $f^{(k)}(c) = 0$ for all k, no longer exists. This constraint has effectively been contained (shifted to) the requirement $f(t) = 0$ for all $t \leq a$. This allows initialization of fractional differential equations.

In summary terminal charging case is backward compatible with integer order calculus and satisfies the applicable criteria established by Ross.

6.7.2 Side Charging

The case for side charging is less definitive. Criteria for backward compatibility is the same as the terminal charging case. Relative to zero property the condition $\psi(f, -p, a, c, t) = \frac{1}{\Gamma(p)} \int_a^c (t-\tau)^{p-1} f(\tau) d\tau = 0 = \psi(h, m, a, c, t)$ is required for side charging since ψ is arbitrary. When these conditions are not met, the zero-order operation on $f(t)$ will return $f(t) + g(t)$, i.e. the original function with extra time function $((g(t))$, the effect of initialization. Relative to linearity, the side charging demands additional requirements about initialization.

These are not so much of an issue as it appears for practical applications. In the solution of fractional differential equations, $\psi(t)$ will be chosen in the much the same manner as initialization are currently chosen for ordinary differential equations in integer order. This will imply the nature of $f(t)$ from a to c. The new aspect is that to achieve a particular initialization for a given composition now requires attention to the initialization of the composing elements.

6.8 The Fundamental Fractional Order Differential Equation

The problem to be addressed in this section is the solution of the fundamental linear fractional order differential equation (6.26). This system is considered to be fundamental because its solution is the fundamental time response, whose combination provides the solution of more complicated systems, analogous to the exponential function for the integer order differential equation. The fundamental equation is

$$_c D_t^q x(t) \equiv {}_c d_t^q x(t) + \psi(x, q, a, c, t) = -ax(t) + bu(t), \quad q > 0 \quad (6.26)$$

where the left side should be interpreted to be the qth derivative of $x(t)$, starting at time c and continuing until time t. Here it will be assumed for clarity that the problem starts at $c = 0$. We also assume temporarily in this section that the initialization function $\psi(x, q, a, c, t) = 0$. Thus we will be concerned only with the forced response for time being. Rewriting (6.26) with these assumptions, we get

$$_0 d_t^q x(t) = -ax(t) + bu(t), \quad q > 0 \quad (6.27)$$

We will use Laplace transform techniques to simplify the solution of this differential equation. In order to do so for this problem, the Laplace transform of the fractional differential is required (given in detail in Chap. 7), ignoring the initialization terms (6.27) can be Laplace transformed as:

$$s^q X(s) = -aX(s) + bU(s) \quad (6.28)$$

6.8 The Fundamental Fractional Order Differential Equation

This equation can be rearranged to obtain the system transfer function:

$$\frac{X(s)}{U(s)} = G(s) = \frac{b}{s^q + a} \quad (6.29)$$

This is the transfer function of the fundamental linear fractional order differential equation. As such, it contains the fundamental "fractional pole" and is the fundamental building block for more complicated fractional order systems. As the constant b in (6.29) is a constant multiplier, it can be assumed without loss of generality to be unity. Typically, transfer functions are used to study various properties of a particular system. Specifically, they can be inverse Laplace transformed to obtain the system impulse response, which can then be used with convolution approaches to the problem. Generally, if $U(s)$ is given, then the product $G(s)U(s)$ can be expanded using partial fractions, and forced response obtained by inverse transforming each separately. To accomplish these tasks, it is necessary to obtain the inverse of (6.29), which is the impulse response, or generalized exponential function, of the fundamental fractional order system.

6.8.1 The Generalized Impulse Response Function

Although the Laplace transform of (6.29) is not contained in standard Laplace transform table, the following transform pair is available:

$$\frac{1}{s^q} = L\left\{\frac{t^{q-1}}{\Gamma(q)}\right\}, \quad q > 0 \quad (6.30)$$

If we expand the right-hand side of (6.29) in describing powers of s, we can then inverse transform the series term by term and obtain the generalized impulse response. Then expanding (6.29) about $s = \infty$, we get

$$G(s) = \frac{1}{s^q + a} = \frac{1}{s^q} - \frac{a}{s^{2q}} + \frac{a^2}{s^{3q}} - \cdots = \frac{1}{s^q} \sum_{n=0}^{\infty} \frac{(-a)^n}{s^{nq}} \quad (6.31)$$

This series can be inverse transformed term by term using (6.30). The result is

$$L^{-1}\{G(s)\} = L^{-1}\left\{\frac{1}{s^q} - \frac{a}{s^{2q}} + \frac{a^2}{s^{3q}} - \cdots\right\} = \frac{t^{q-1}}{\Gamma(q)} - \frac{at^{2q-1}}{\Gamma(2q)} + \frac{a^2 t^{3q-1}}{\Gamma(3q)} - \cdots \quad (6.32)$$

The right side of (6.32) can now be collected into summation and used as definition of the generalized impulse response function:

$$F_q[-a, t] \equiv t^{q-1} \sum_{n=0}^{\infty} \frac{(-a)^n t^{nq}}{\Gamma(\{n+1\}q)}, \quad q > 0 \quad (6.33)$$

We thus have important Laplace identity as $L\{F_q[a,t]\} \leftrightarrow \frac{1}{s^q-a}, q > 0$. Also the F function is generalization of exponential function for $q = 1$,

$$F_1[-a,t] = \sum_{n=0}^{\infty} \frac{(-at)^n}{\Gamma(n+1)} = e^{-at}$$

This generalization is the basis for the solution of most linear fractional order differential equations.

Here we have established the F function as the impulse response of the fundamental linear differential equations. This function is important because it will allow the creation of concise theory for fractional order systems, which is a generalization of that of integer order systems, and where the F function generalizes and replaces the usual exponential function.

Several other variants of this F function is possible for the solution of the fundamental equation (6.27), like Miller–Ross function, R function, and G function, listed in Chap. 2.

From (6.32) and (6.33), we obtained impulse response as:

$$g(t) = L^{-1}\{G(s)\} = t^{q-1} \sum_{n=0}^{\infty} \frac{(-a)^n t^{nq}}{\Gamma(nq+q)} \equiv F_q[-a,t], \quad q > 0 \qquad (6.34)$$

The function $F_q[a,t]$ is closely related to the Mittag-Leffler function $E_q[at^q]$, where one-parameter Mittag-Leffler function in series form is defined as:

$$E_q[x] \equiv \sum_{n=0}^{\infty} \frac{x^n}{\Gamma(nq+1)}, \quad q > 0 \qquad (6.35)$$

Letting $x = -at^q$, this becomes

$$E_q[-at^q] \equiv \sum_{n=0}^{\infty} \frac{(-a)^n t^{nq}}{\Gamma(nq+1)}, \quad q > 0 \qquad (6.36)$$

which is similar to $F_q[-a,t]$ expressed in (6.34), but not same as (6.34).

The Laplace transform of this Mittag-Leffler function (6.36) can also be obtained via term by term transform of the series expansion as:

$$L\{E_q[-at^q]\} = L\left\{\frac{1}{\Gamma(1)} - \frac{at^q}{\Gamma(1+q)} + \frac{a^2 t^{2q}}{\Gamma(1+2q)} + \cdots\right\}$$
$$= \frac{1}{s} - \frac{a}{s^{q+1}} + \frac{a^2}{s^{2q+1}} + \cdots \qquad (6.37)$$

6.8 The Fundamental Fractional Order Differential Equation

or compactly:

$$L\{E_q[-at^q]\} = \frac{1}{s}\left[1 - \frac{1}{s^q} + \frac{a^2}{s^{2q}} + \ldots\right] = \frac{1}{s}\sum_{n=0}^{\infty}\left(\frac{-a}{s^q}\right)^n = \frac{1}{s}\sum_{n=0}^{\infty}\frac{(-a)^n}{s^{nq}} \quad (6.38)$$

It should be noted that the summation expression (6.38) is similar to (6.31). Using (6.31), (6.38) can be rewritten as:

$$L\{E_q[-at^q]\} = \frac{1}{s}\left[\frac{s^q}{s^q + a}\right] \quad (6.39)$$

or equivalently:

$$L\{E_q[-at^q]\} = \frac{1}{s}\left[s^q L\{F_q[-a,t]\}\right] \quad (6.40)$$

Thus a general result can be expressed as:

$$L\{E_q[\pm at^q]\} = \frac{s^{q-1}}{s^q \mp a}, \quad q > 0 \quad (6.41)$$

From (6.40), the relation between Mittag-Leffler (E_q) function and Robotnov–Hartley function (F_q) is

$$_0d_t^{q-1} F_q[a,t] = E_q[at^q] \quad (6.42)$$

This demonstration was to show a method of obtaining solution of the "fundamental fractional order (linear) differential equation" that is (6.27) by use of Robotnov–Hartley function. This F function was utilized by Robotnov to study hereditary integrals in solid mechanics. Solution to (6.27) may be obtained in terms of Miller–Ross function and its Laplace transform. Miller–Ross function is the fractional derivative of the exponential function defined as

$E_t(v,a) \equiv {_0d_t^{-v}}\exp(at)$, whose Laplace transform is $L\{E_t(v,a)\} = \frac{s^{-v}}{s-a}$.

Also recent developments to study diffusion and fractional kinetic equations use more complicated Fox functions in solving of fractional order differential equations.

Extending this developed technique, we obtain the solution of (6.27) for a unit step input excitation. This can be obtained via Laplace transforms by transforming the input function $u(t)$, its Laplace is $1/s$, which must be multiplied by the transfer function $G(s)$, (6.29), where $b = 1$ is taken. We get

$$X(s) = \frac{1}{s}\left[\frac{1}{s^q + a}\right] \quad (6.43)$$

Manipulating (6.43), we obtain

$$X(s) = \frac{1/a}{s}\left[\frac{a}{s^q+a}\right] = \frac{1/a}{s}\left[1 - \frac{s^q}{s^q+a}\right] = \frac{1/a}{s} - \frac{s^q/a}{s(s^q+a)} \quad (6.44)$$

Using expression (6.40), (6.44) is inverse Laplace transformed; the result is the step response of the system:

$$x(t) = L^{-1}\left\{\frac{1}{s(s^q+a)}\right\} = \frac{1}{a}\left(1 - E_q\left[-at^q\right]\right) = \frac{1}{a}\left(h(t) - E_q\left[-at^q\right]\right) \quad (6.45)$$

Heaviside step is $h(t) = \begin{cases} 1 & t \geq 0 \\ 0 & t < 0 \end{cases}$

Taking integer derivative of (6.45) gives $F_q[-a, t]$, the impulse response:

$$F_q[-a, t] = \frac{1}{a}\frac{d}{dt}\left(h(t) - E_q\left[-at^q\right]\right) \quad (6.46)$$

Referring to (6.40) and multiplying the Laplace transforms, thereby s^{-q} gives

$$s^{-q} L\left\{E_q\left[-at^q\right]\right\} = s^{-1} L\left\{F_q[-a, t]\right\} = \frac{1}{s(s^q+a)},$$

which is (6.43). Inverse transforming this equation using expression (6.45) shows that the step response of (6.27) with $b = 1$ is also the qth fractional integral of the Mittag-Leffler function, that is:

$$x(t) = L^{-1}\left\{\frac{1}{s(s^q+a)}\right\} = \frac{1}{a}\left[h(t) - E_q(-at^q)\right] = {}_0d_t^{-q} E_q\left[-at^q\right] \quad (6.47)$$

Few more interesting Laplace pairs can be obtained by taking the qth derivative of the F function. Taking the un-initialized fractional derivative (${}_0d_t^q$), and in Laplace domain multiplying by s^q gives

$$L^{-1}\left\{\frac{s^q}{s^q+a}\right\} = {}_0d_t^q F_q[-a, t] \quad (6.48)$$

This expression (6.48) is also the integer derivative of the Mittag-Leffler function:

$$L^{-1}\left\{\frac{s^q}{s^q+a}\right\} = {}_0d_t^q F_q[-a, t] = {}_0d_t^q E_q\left[-at^q\right] \quad (6.49)$$

The expression (6.49) can also be rewritten as:

$$L^{-1}\left\{\frac{s^q}{s^q+a}\right\} = L^{-1}\left\{1 - \frac{a}{s^q+a}\right\} = \delta(t) - a F_q[-a, t] \quad (6.50)$$

Observing the expressions (6.48) and (6.50), we can write the following:

$${}_0d_t^q F_q[-a, t] = \delta(t) - a F_q[-a, t] \tag{6.51}$$

This $F_q[a.t]$ function is generalization of the exponential function of the integer order calculus where it demonstrates the "eigenfunction property"; i.e., returning the same function upon the qth fractional differentiation. Also (6.51) shows that $x(t) = F_q[-a, t]$, Robotnov–Hartley function is impulse response of the fundamental fractional order differential equation given by (6.27), for $u(t) = \delta(t)$ and $b = 1$.

6.9 Concluding Comments

Strong motivation exists for the study and development of the fractional calculus. This may be readily verified and validated by a large number of applications discussed in the preceding chapters, where need for fractional calculus and initialization in particular is pointed. Fractional integration and fractional differentiation processes require the use of non-constant initialization function. This initialization function is the behavior of the function before the process of differintegration is taken up, and makes up the process continuous in future. The initialization function is history or memory associated with the process of differintegration of the function since its birth. This history or memory fades away as time passes by and appears as output of fractional differintegrator as an added function. The solution of fundamental fractional differential equation contains fractional pole in its transfer function and property of which is important in system analysis. The initialized differintegration process is the generalization of the total calculus theory, for fractional order systems or integer order systems. The unifying concepts and notation of fractional calculus provide a significant benefit that greatly simplifies the solution of certain differential equations (distributed systems). Perhaps the strongest motivation to develop the fractional calculus is the belief that a wide variety of physical problems that have resisted compact (and first principles) description when using integer order calculus will yield to the methods of fractional calculus, otherwise major recourse was the probabilistic methods.

Chapter 7
Generalized Laplace Transform for Fractional Differintegrals

7.1 Introduction

Differential equations of fractional order appear more and more frequently in various research areas of science and engineering. An effective method for solving such equations is needed. The method of Laplace transforms technique gives almost a unified approach to solve the fractional differential equations. Also generalization of the same in view of initial conditions appropriately put (terminal/side charging) gives unified generalized approach. In this chapter, scalar initialization and vector initialization problem is taken to describe the approaches developed for initialization function. These problems give insight into fractional "state" variable concepts and general system description of fractional order systems, and controls. For fractional order control system, stability analysis transformed Laplace $s^q \to w$ plane (wedge) is introduced. The pole placement and its properties for control system stability for fractional order systems are carried on in this w-plane. The realization of fractional Laplace operator by rational function approximation is also introduced in this chapter.

7.2 Recalling Laplace Transform Fundamentals

Let us recall some basic facts about Laplace transforms, developed for classical integer order calculus. The function $F(s)$ of complex variable (frequency) s is defined by:

$$F(s) = L\{f(t)\} = \int_0^\infty e^{-st} f(t) dt$$

this is called Laplace transform of the function $f(t)$, which is called the original. For Laplace transform to exist, $f(t)$ must be of an exponential order α. In other words, the function $f(t)$ must not grow faster than a certain exponential function when t tends to infinity. The original $f(t)$ can be restored from Laplace transform $F(s)$ with the help of inverse transform:

$$f(t) = L^{-1}\{F(s)\} = \frac{1}{j2\pi} \int_{\gamma-j\infty}^{\gamma+j\infty} e^{st} F(s) ds$$

where $\gamma = 0$ for all singularities of $F(s)$ in the left half of s-plane (LHP) i.e. all $F(s)$ poles at LHP.

Laplace transform of $\int_b^t f(t)dt$ is given by

$$L\left\{\int_b^t f(t)dt\right\} = \frac{1}{s}F(s) + \frac{1}{s}\int_b^0 f(t)dt.$$

Constructing this for initialized case, the generalized integer order Laplace transformation for integration operation can be expressed as:

$$L\left\{{}_bD_t^{-1}f(t)\right\} = L\left\{{}_bd_t^{-1}f(t) + \psi(f,-1,a,b,t)\right\}$$

$$= L\left\{\int_b^t f(t) + \psi(f,-1,a,b,t)\right\}$$

$$L\left\{{}_bD_t^{-1}f(t)\right\} = \frac{1}{s}F(s) + \frac{1}{s}\int_b^0 f(t)dt + L\{\psi(f,-1,a,b,t)\}.$$

However, in the most general case, the ψ is selected arbitrarily, if chosen as constant K, then since $L\{K\} = K/s$, it is clear that this term contains initialization effect of the second term on the RHS of the above Laplace expression for initialized integer order integration. Hence, it is not necessary to include such terms that redundantly bring in the effect of initialization from the integer order calculus. Therefore:

$$L\left\{{}_bD_t^{-1}f(t)\right\} = \frac{1}{s}F(s) + L\{\psi(f,-1,a,b,t)\}$$

For multiple integer order integrals, the Laplace transform derivation is as follows:

$$L\{g(t)\} = L\left\{\int_b^t \int_b^{t_1} \int_b^{t_2} \ldots \int_b^{t_{n-1}} f(t_n) dt_n dt_{n-1} \ldots dt_2 dt_1\right\};$$ this is to be evaluated.

For convenience, write $g(t_j) = \int_b^{t_j} g(t_{j+1}) dt_{j+1}$, for $j = 1, 2, \ldots n-2$, and $g(t_{n-1}) = \int_b^{t_{n-1}} f(t_n) dt_n$. Then starting from out side, we have

$$L\{g(t)\} = L\left\{\int_b^t g(t_1) dt_1\right\} = \frac{1}{s}L\{g(t_1)\} + \frac{1}{s}\int_b^0 g(t_1) dt_1 = \frac{1}{s}L\{g(t_1)\} + \frac{c_1}{s},$$

where constant $c_1 = \int_b^0 g(t_1)dt_1$. Going one level inside, i.e. replacing the $g(t_1)$ with $\int_b^{t_1} g(t_2)dt_2$, we get

$$L\{g(t)\} = \frac{1}{s}\left[L\left\{\int_b^{t_1} g(t_2)dt_2\right\}\right] + \frac{c_1}{s} = \frac{1}{s}\left[\frac{1}{s}L\{g(t_2)\} + \frac{c_2}{s}\right] + \frac{c_1}{s}$$

$$= \frac{1}{s^2}L\{g(t_2)\} + \frac{c_2}{s^2} + \frac{c_1}{s}$$

Repeating this n times, we get the result as:

$$L\left\{\int_b^t \int_b^{t_1} \int_b^{t_2} \cdots \int_b^{t_{n-1}} f(t_n)dt_n dt_{n-1} \ldots dt_2 dt_1\right\}$$

$$= \frac{1}{s^n}F(s) + \sum_{i=1}^n \frac{c_i}{s^i}, n = 1, 2, 3, \ldots,$$

where coefficients c_i is given as $c_i = \int_b^0 g(t_i)dt_i$.

Consider the integer order derivative and its Laplace transform as $L\left\{\frac{d}{dt}f(t)\right\} = L\{f^1(t)\} = sF(s) - f(0^+)$. For the initialized calculus case, the formulation is $L\{_0D_t^1 f(t)\} = L\left\{\frac{df(t)}{dt} + \psi(f, 1, a, 0, t)\right\}$. Substituting the first definition into the second one we get $L\{_0D_t^1 f(t)\} = sF(s) - f(0^+) + L\{\psi(f, 1, a, 0, t)\}$. Again in the most general case, the ψ is selected arbitrarily. If it is chosen as value of the function $f(t)$ at $t = 0^+$, represented by Dirac delta function as $\psi = -f(t)\delta(t - 0^+)$, then $L\{\psi\} = L\{-f(t)\delta(t - 0^+)\} = -f(0^+)$. This term then contains the initialization effect brought in by the integer order calculus Laplace expansion. It is again not necessary to include the redundant term.

The notation $f(0^+)$ may be changed to $f(0)$ with the understanding that $t = 0$ will be actually $t = 0^+$. For repeated derivative, the Laplace transform gives

$$L\{f^{(n)}(t)\} = s^n F(s) - s^{n-1} f(0) - s^{n-2} f^{(1)}(0) - \ldots - f^{(n-1)}(0)$$

7.3 Laplace Transform of Fractional Integrals

Consider starting point of integration as $c = 0$, for simplicity. The Laplace transform of the initialized fractional integral looks like

$$L\{_0D_t^{-q} f(t)\} = \int_0^\infty e^{-st}\left(\int_0^t \frac{(t-\tau)^{q-1}}{\Gamma(q)} f(\tau)d\tau + \psi(f, -q, a, 0, t)\right)dt \quad (7.1)$$

$q > 0, t > 0$. Apply convolution for Laplace transforms, i.e.,

$$h(t)^*g(t) = \int_0^t h(\tau)g(t-\tau)d\tau \leftrightarrow H(s)G(s).$$

For the above fractional integral, take $h(t) = f(t).\&.g(t) = \frac{t^{q-1}}{\Gamma(q)} \leftrightarrow \frac{1}{s^q}$.
The Laplace of the fractional integral therefore is $F(s)G(s) = \frac{1}{s^q}L\{f(t)\}, q > 0$.
This gives the result as:

$$L\left\{{}_0D_t^{-q}f(t)\right\} = \frac{1}{s^q}L\{f(t)\} + L\{\psi(f,-q,a,0,t)\}, q > 0 \quad (7.2)$$

Here the observation is that the initialization function $\psi(f,-q,a,0,t)$ may be thought to have equivalent (compound) effect of initialization to create ${}_0D_t^{-q}f(t)$. For fractional case, ${}_0D_t^{-q}f(t)$ can be composed in infinite ways as opposed to integer order calculus, where only, integer order combinations are possible for composition, as indicated below for integer order n:

$L\{f^{(n)}(t)\} = s^n F(s) - s^{n-1}f(0) - s^{n-2}f^{(1)}(0)\ldots + f^{(n-1)}(0)$. This implies that $f^{(n)}$ is decomposed into n separate differentiations of unity order; conversely, $f^{(n)}$ is composed of n separate discrete differentiation. But for fractional order, this composition is not discrete but can thus be decomposed in infinite ways. This is described next.

7.3.1 Decomposition of Fractional Integral in Integer Order

For generalized integration (with initialization), the composition law holds. The decomposition of the following Laplace of initialization function is demonstrated by integration by parts, for $q > 1$. The initialization function $\psi(f,-q,a,0,t)$ is qth order integration of $f(t)$ from a to 0 and is function of t, defined as in the Chap. 6 of initialized differintegrals.

$$\psi(f,-q,a,0,t) = \frac{1}{\Gamma(q)}\int_a^0 (t-\tau)^{q-1}f(\tau)d\tau \quad q > 1 \quad (7.3)$$

$$L\{\psi(f,-q,a,0,t)\} = \int_0^\infty e^{-st}\psi(f,-q,a,0,t)dt \quad (7.4)$$

Take $u = \psi(f,-q,a,0,t), dv = e^{-st}dt$, then
$du = \frac{d}{dt}\psi(f,-q,a,0,t) = \psi^{(1)}$, and $v = \frac{e^{-st}}{-s}$ yields

7.3 Laplace Transform of Fractional Integrals

$$L\{\psi(f,-q,a,0,t)\} = \left[\frac{e^{-st}}{-s}\psi(f,-q,a,0,t)\right]_0^\infty$$
$$+ \frac{1}{s}\int_0^\infty e^{-st}\psi^{(1)}(f,-q,a,0,t)dt \qquad (7.5)$$

$$L\{\psi(f,-q,a,0,t)\} = 0 + \frac{1}{s}\psi(f,-q,a,0,t)_{at\,t=0} + \frac{1}{s}L\{\psi^{(1)}(f,-q,a,0,t)\} \qquad (7.6)$$

Repeating the same another time gives

$$L\{\psi(f,-q,a,0,t)\} = \frac{1}{s}\psi(f,-q,a,0,t)_{at\,t=0} + \frac{1}{s}\left[\frac{1}{s}\psi^{(1)}(f,-q,a,0,t)_{at\,t=0}\right.$$
$$\left. + \frac{1}{s}L\{\psi^{(2)}(f,-q,a,0,t)\}\right] \qquad (7.7)$$

Now repeating the process a total number of n times, where n is integer, such that $n+1 > q > n$ gives the expression for the equivalent initialization function. For $q = 2.3 > 1$, the $n = 2$.

$$L\{\psi(f,-q,a,0,t)\} = \frac{1}{s^n}L\{\psi^{(n)}(f,-q,a,0,t)\}$$
$$+ \sum_{j=1}^{n}\frac{1}{s^j}\psi^{(j-1)}(f,-q,a,0,t)_{at\,t=0} \qquad (7.8)$$

Putting this the Laplace transform of the fractional integral yields

$$L\{_0D_t^{-q}f(t)\} = \frac{1}{s^q}L\{f(t)\} + \frac{1}{s^n}L\{\psi^{(n)}(f,-q,a,0,t)\}$$
$$+ \sum_{j=1}^{n}\frac{1}{s^j}\psi^{(j-1)}(f,-q,a,0,t)_{at\,t=0}. \qquad (7.9)$$

The expression states that the qth differintegral is composed (or can be decomposed) of n order 1 integer integrations and a fractional integration of order $q - n$, refer Fig. 7.1. For $q = 2.3$, meaning the composition have $n = 2$, full integer order integration, and $q - n = 0.3$, order fractional integration.

Further more, the order 1 integrations are each initialized by a constant $\psi^{(j-1)}_{at\,t=0}$. Following, we describe for sake of comparison the Laplace transform for multiple integer order integrals. The compatibility is observed for (7.9) and (7.10), for $q = n = 1, 2, 3\ldots$ and properly choosing ψ.

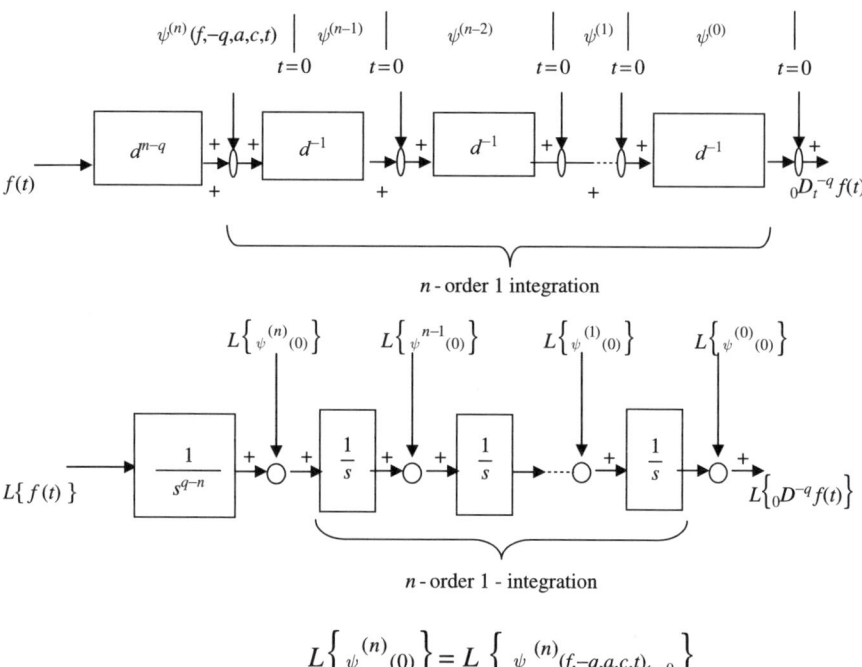

Decomposition for $q > 1$

Fig. 7.1 Block diagram representing integer order decomposition of fractional integral (time and frequency domain)

$$\begin{aligned}
L\{g(t)\} &= L\left\{\int_b^t \int_b^{t_1} \int_b^{t_2} \int \ldots\ldots \int_b^{t_{n-1}} \int f(t_n) dt_n dt_{n-1}\ldots\ldots dt_2 dt_1\right\} \\
&= \frac{1}{s^n} L\{f(t)\} + \frac{c_1}{s} + \frac{c_2}{s^2} + \ldots + \frac{c_n}{s^n} \\
&= \frac{1}{s^n} L\{f(t)\} + \sum_{j=1}^{n} \frac{c_j}{s^j}
\end{aligned} \quad (7.10)$$

The coefficients are $c_j = \int_b^0 g(t_j) dt_j$. In Fig. 7.1, the integer order integrations could just be replaced by general (fractional) integrations, each also of order 1. Each of these then will allow non-constant initialization function. Figure 7.2 is ${}_0D_t^{-q} f(t)$ decomposed into n generalized order 1 integrations and a fractional integration of order $-q + n$, where n is the greatest integer less than q. The breakup expression for Fig. 7.2 is

$$\begin{aligned}
{}_0D_t^{-q} f(t) = x_n(t) &= \psi_1 + {}_0d_t^{-1} x_{n-1}(t) = \psi_1 + {}_0d_t^{-1}(\psi_2 + {}_0d_t^{-1} x_{n-2}(t)) \\
&= \psi_1 + {}_0d_t^{-1}\psi_2 + {}_0d_t^{-2}(\psi_3 + {}_0d_t^{-1} x_{n-3}(t)) \\
&= \psi_1 + {}_0d_t^{-1}\psi_2 + {}_0d_t^{-2}\psi_3 + {}_0d_t^{-3} x_{n-3}(t)
\end{aligned} \quad (7.11)$$

7.3 Laplace Transform of Fractional Integrals

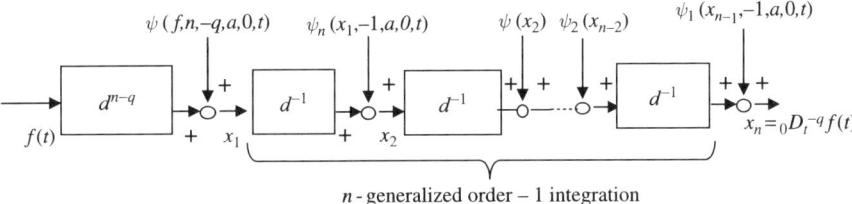

Fig. 7.2 Block diagram of integer order decomposition of fractional integration of order q, for $t > c = 0$

Observing the fact that $x_1(t) = {}_0D_t^{-q+n} f(t) = \psi(f, -q+n, a, 0, t) + {}_0d_t^{-q+n} f(t)$, repeating the above iteration n times, we have the following:

$${}_0D_t^{-q} f(t) = {}_0d_t^{-q} f(t) + {}_0d_t^{-n} \psi(f, -q+n, a, 0, t)$$
$$+ \sum_{j=1}^{n} {}_0d_t^{-(j-1)} \psi_j(x_j, -1, a, 0, t) \qquad (7.12)$$

as the mathematical representation of Fig. 7.2. Taking the Laplace transforms, we get

$$L\left\{{}_0D_t^{-q} f(t)\right\} = \frac{1}{s^q} L\{f(t)\} + \frac{1}{s^n} L\{\psi(f, -q+n, a, 0, t)\}$$
$$+ \sum_{j=1}^{n} \frac{1}{s^{(j-1)}} L\{\psi_j(x_j, -1, a, 0, t)\} \qquad (7.13)$$

This expression is generalization of the expressions of Fig. 7.1, where the initialization effect was done by constants, values of $\psi^{(n)}(t)_{at\ t=0}$. In this case, the initialization effect is carried out by functions of time instead (i.e. ψ_n). Here for $q = 2.3$, the integer $n = 2$, meaning the composition with three integer order integration and one fractional order $-q + n = -0.3$ integration.

7.3.2 Decomposition of Fractional Order Integral in Fractional Order

Here the decomposition of ${}_0D_t^{-q} f(t)$ is not limited to only integer order integral elements. Refer Fig. 7.3, where this decomposition is indicated. The mathematics is explained below (for convenience subscript is dropped i.e. ${}_0D_t^{-q} \to D^{-q}$)

$${}_0D_t^{-q} f(t) = x_{n+1}(t) = D^{-q_n} D^{-q_{n-1}} \ldots D^{-q_2} D^{-q_1} x_1(t) \quad t > 0. \& .q_j > 0 \qquad (7.14)$$

where $q = \sum_{i=1}^{n} q_i$

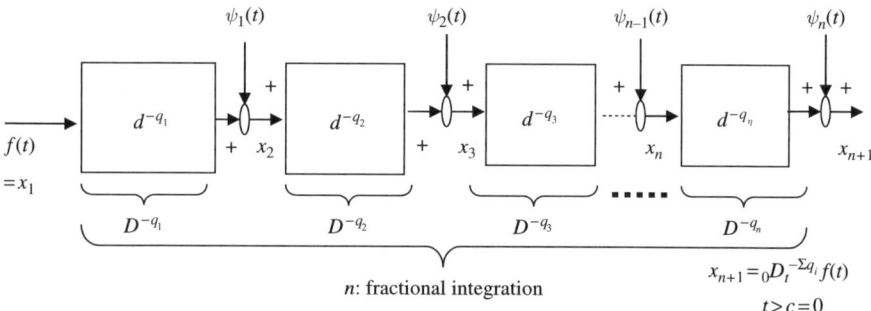

Fig. 7.3 Decomposition of fractional order integral in fractional order

Then starting from inside, $L\{x_2(t)\} = L\{_0 d_t^{q_1} x_1(t)\} + L\{\psi_1(x_1, -q_1, a, 0, t)\}$ can be written with simplified symbols as

$$L\{x_2\} = L\{d^{-q_1} x_1\} + \psi_1(s) = s^{-q_1} L\{x_1\} + \psi_1(s)$$

$$L\{x_3\} = s^{-q_2}\{s^{-q_1} L\{x_1\} + \psi_1(s)\} + \psi_2(s) = s^{-q_2-q_1} L\{x_1\} + s^{-q_1}\psi_1(s) + \psi_2(s)$$
$$L\{x_4\} = s^{-q_3-q_2-q_1} L\{x_1\} + s^{-q_3-q_2}\psi_1(s) + s^{-q_3}\psi_2(s) + \psi_3(s).$$

Repeating this process till x_{n+1} and defining B_a as $B_a = \sum_{i=a}^{n} q_i$, we can have the general form of decomposition expression as, where $1 \leq a \leq n$,

$$L\{x_{n+1}\} = s^{-B_1} L\{x_1\} + s^{-B_2}\psi_1(s) + s^{-B_3}\psi_2(s) + \ldots + s^{-B_n}\psi_{n-1}(s) + \psi_n(s) \quad (7.15)$$

Summarizing this in compact form we get

$$L\{_0 D_t^{-q} f(t)\} = s^{-q} L\{f(t)\} + \psi_n(s) + \sum_{j=1}^{n-1} s^{-B_{j+1}} \psi_j(s)$$

$$B_a = \sum_{i=a}^{n} q_i \quad 1 \leq a \leq n \quad (7.16)$$

The effective initialization here in this case is $L\{\psi_{eff}\} = \psi_n(s) + \sum_{j=1}^{n-1} s^{-B_{j+1}} \psi_j(s)$

7.4 Laplace Transformation of Fractional Derivatives

The derivative starting point is taken as $c = 0$ for simplicity, and the function $f(t)$ is born at a, before $t = a$ the function is zero. Here, we will find $L\{_0 D_t^u f(t)\} = L\{_0 D_t^m {_0 D_t^{-p}} f(t)\}$ with $u > 0$, and m is the least integer greater

7.4 Laplace Transformation of Fractional Derivatives

than u, such that $u = m - p$. For $u = 2.3$, the integer $m = 3$. In the case of integration decomposition, the integer n is chosen as the greatest integer less than the fractional order of integration q. Using the definitions and expanding we get

$$L\left\{_0D_t^u f(t)\right\} = L\left\{_0D_t^m {_0D_t^{-p}} f(t)\right\}$$

$$= L\left\{\frac{d^m}{dt^m}\left(\int_0^t \frac{(t-\tau)^{p-1}}{\Gamma(p)} f(\tau)d\tau\right)\right\}$$

$$+ L\left\{\frac{d^m}{dt^m}\psi(f,-p,a,0,t)\right\} + L\left\{\psi(h,m,a,0,t)\right\} \quad (7.17)$$

where $h(t) = {_0d_t^{-p}} f(t)$. Defining equivalent initialization function as $\psi_{eq}(f, u, a, 0, t)$:

$$\psi_{eq} = \left\{\frac{d^m}{dt^m}\psi_{eq}(f,-p,a,0,t)\right\} + \left\{\psi(h,m,a,0,t)\right\} \quad (7.18)$$

Now consider the first term of the expanded expression i.e.

$$L\left\{\frac{d^m}{dt^m}\left(\int_0^t \frac{(t-\tau)^{p-1}}{\Gamma(p)} f(\tau)d\tau\right)\right\} = \int_0^\infty e^{-st} \frac{d}{dt}\left(\frac{d^{m-1}}{dt^{m-1}}\int_0^t \frac{(t-\tau)^{p-1}}{\Gamma(p)} f(\tau)d\tau\right) dt \quad (7.19)$$

Apply integration by parts by selecting $u = e^{-st}$ and

$$dv = \frac{d}{dt}\left(\frac{d^{m-1}}{dt^{m-1}}\int_0^t \frac{(t-\tau)^{p-1}}{\Gamma(p)} f(\tau)d\tau\right) dt$$

Therefore, $du = -se^{-st} dt$ and $v = \frac{d^{m-1}}{dt^{m-1}}\int_0^t \frac{(t-\tau)^{p-1}}{\Gamma(p)} f(\tau)d\tau$ yields

$$= e^{-st}\frac{d^{m-1}}{dt^{m-1}}\int_0^t \frac{(t-\tau)^{p-1}}{\Gamma(p)} f(\tau)d\tau\Big]_0^\infty + s\int_0^\infty e^{-st}\frac{d}{dt}\left(\frac{d^{m-2}}{dt^{m-2}}\int_0^t \frac{(t-\tau)^{p-1}}{\Gamma(p)} f(\tau)d\tau\right) dt \quad (7.20)$$

The first term is zero after putting the end values; therefore, we get the Laplace expression as

$$L\left\{\frac{d^m}{dt^m}\left(\int_0^t \frac{(t-\tau)^{p-1}}{\Gamma(p)} f(\tau)d\tau\right)\right\} = sL\left\{\frac{d}{dt}\left(\frac{d^{m-2}}{dt^{m-2}}\int_0^t \frac{(t-\tau)^{p-1}}{\Gamma(p)} f(\tau)d\tau\right)\right\} \quad (7.21)$$

Repeating the process m times yields

$$L\left\{\frac{d^m}{dt^m}\int_0^t \frac{(t-\tau)^{p-1}}{\Gamma(p)} f(\tau)d\tau\right\} = s^m L\left\{_0d_t^{-p} f(t)\right\} \qquad (7.22)$$

using this result, the Laplace expression for fractional derivative is

$$L\left\{_0D_t^u f(t)\right\} = L\left\{_0D_t^m {_0D_t^{-p}} f(t)\right\}$$
$$= s^m L\left\{_0d_t^{-p} f(t)\right\} + L\left\{\frac{d^m}{dt^m}\psi(f,-p,a,0,t)\right\}$$
$$+L\{\psi(h,m,a,0,t)\} \qquad (7.23)$$

or $L\left\{_0D_t^u f(t)\right\} = s^m L\left\{_0d_t^{-p} f(t)\right\} + L\{\psi_{eq}(f,u,a,0,t)\}$ (7.24)

here applying integral Laplace result for Laplace of $_0d_t^{-p}$, we get

$$L\left\{_0D_t^u f(t)\right\} = s^{m-p} L\{f(t)\} + L\{\psi_{eq}(f,u,a,0,t)\}$$
$$= s^u L\{f(t)\} + L\{\psi_{eq}(f,u,a,0,t)\} \qquad (7.25)$$

This is the most general form of Laplace transform of the fractional derivative and is similar to what is obtained for Laplace of fractional integration. As the case with fractional integrals, the fractional derivatives also can be decomposed (or is composed off) infinite ways, thus several possible formulations exist for $\psi_{eq}(f,u,a,c,t)$.

7.4.1 Decomposition of Fractional Order Derivative in Integer Order

Equivalent form for $L\{\psi(f,u,a,0,t)\}$ of (7.25) is considered. Consider the expression that appeared in the previous derivation (7.17) i.e.

$$L\left\{\frac{d^m}{dt^m}\psi(f,-p,a,0,t)\right\} = \int_0^\infty e^{-st}\frac{d}{dt}\left(\frac{d^{m-1}}{dt^{m-1}}\psi(f,-p,a,0,t)\right)dt$$

For integrating by parts, take $u = e^{-st}$ and $dv = \frac{d}{dt}\frac{d^{m-1}}{dt^{m-1}}\psi(f,-p,a,0,t)dt$, so $du = -se^{-st}dt$ and $v = \frac{d^{m-1}}{dt^{m-1}}\psi(f,-p,a,0,t)$. Therefore,

7.4 Laplace Transformation of Fractional Derivatives

$$L\left\{\frac{d^m}{dt^m}\psi(f,-p,a,0,t)\right\} = e^{-st}\frac{d^{m-1}}{dt^{m-1}}\psi(f,-p,a,0,t)]_0^\infty$$

$$+ s\int_0^\infty e^{-st}\frac{d^{m-1}}{dt^{m-1}}\psi(f,-p,a,0,t)dt$$

$$= -\frac{d^{m-1}}{dt^{m-1}}\psi(f,-p,a,0,t)]_{t=0}$$

$$+ sL\left\{\frac{d^{m-1}}{dt^{m-1}}\psi(f,-p,a,0,t)\right\} \quad (7.26)$$

This result is thus used to get

$$L\left\{\frac{d}{dt}\frac{d^{m-2}}{dt^{m-2}}\psi(f,-p,a,0,t)\right\} = -\left(\frac{d^{m-2}}{dt^{m-2}}\psi(f,-p,a,0,t)\right)_{t=0}$$

$$+ sL\left\{\frac{d^{m-2}}{dt^{m-2}}\psi(f,-p,a,0,t)\right\} \quad (7.27)$$

Therefore,

$$L\left\{\frac{d^m}{dt^m}\psi(f,-p,a,0,t)\right\} = -\frac{d^{m-1}}{dt^{m-1}}\psi(f,-p,a,0,t)]_{t=0}$$

$$+ s\left[-\left(\frac{d^{m-2}}{dt^{m-2}}\psi(f,-p,a,0,t)\right)_{t=0} + sL\left\{\frac{d^{m-2}}{dt^{m-2}}\psi(f,-p,a,0,t)\right\}\right] \quad (7.28)$$

Repeating this process m times and writing

$$\frac{d^k}{dt^k}\psi(f,-p,a,0,t) = \psi^{(k)}(f,-p,a,0,t)$$

gives

$$L\left\{\frac{d^m}{dt^m}\psi(f,-p,a,0,t)\right\} = s^m L\{\psi(f,-p,a,0,t)\}$$

$$- \sum_{j=1}^{m} s^{j-1}\psi^{m-j}(f,-p,a,0,t)]_{t=0} \quad (7.29)$$

Substituting the above obtained expression into equation for Laplace transform of fractional derivative, we obtain Laplace transform of generalized derivative decomposed into integer differentiations with $\psi^{m-j}(f,-p,a,c,t)_{at\ t=0}$, a constant initialization, as:

$$L\left\{_0D_t^u f(t)\right\} = L\left\{_0D_t^m {}_0D_t^{-p} f(t)\right\} = s^m L\left\{_0d_t^{-p} f(t)\right\} + s^m L\left\{\psi(f, -p, a, 0, t)\right\}$$

$$-\sum_{j=1}^m s^{j-1}[\psi^{m-j}(f, -p, a, 0, t)]_{t=0} + L\{\psi(h, m, a, 0, t)\} \quad (7.30)$$

Here

$$_0d_t^{-p} f(t) = \frac{1}{\Gamma(p)} \int_0^t (t-\tau)^{p-1} f(\tau) d\tau = h(t),$$

the un-initialized fractional integration starting at $t = c = 0$.

$$L\left\{_0D_t^u f(t)\right\} = L\left\{_0D_t^m {}_0D_t^{-1} f(t)\right\}$$

$$= s^u L\{f(t)\} - \sum_{j=1}^m s^{j-1}(\psi^{m-j}(f, -p, a, 0, t)]_{t=0}$$

$$+ L\{\psi(h, m, a, 0, t)\} \quad (7.31)$$

Figure 7.4 shows the decomposition.

Note what has been done here is to decompose the integer order derivative part of $_0D_t^u f(t)$ namely into m integer order 1 differentiations. For integer order Laplace transform of derivative, we have $L\left\{f^{(m)}(t)\right\} = s^m L\{f(t)\} - \sum_{j=1}^m s^{j-1} f^{(m-j)}(0)$. In (7.31) put $p=0$. This becomes zero-order operation, then set $\psi(h, m, a, 0, t) = 0$ as discussed for terminal charging case for integer order initialization, in Chap. 6, gives a specialized case with $\psi^{(m-j)}(f, 0, a, 0, t)_{at\ t=0} = f^{(m-j)}(0)$. This will give the above result of integer order repeated derivative Laplace relation as indicated above.

In context of (7.31), the Laplace expression of total initialized system is

$$L\{\psi(f, u, a, 0, t)\} = -\sum_{j=1}^m s^{j-1}[\psi^{(m-j)}(f, -p, a, 0, t)]_{at\ t=0} + L\{\psi(h, m, a, 0, t)\}.$$

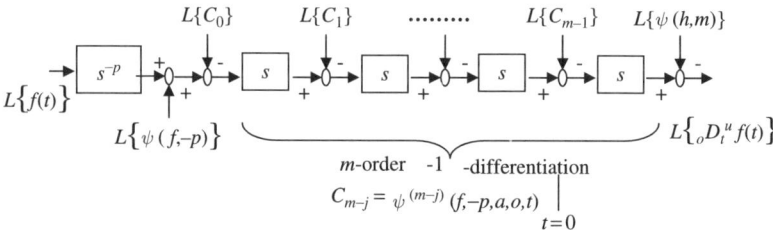

Fig. 7.4 Decomposition of fractional derivative into integer order

7.4 Laplace Transformation of Fractional Derivatives

Further generalization is possible with (7.31). From (7.17) consider the term $L\left\{\frac{d^m}{dt^m}\psi(f,-p,a,0,t)\right\}$ and for complete arbitrary initialization let

$$\psi(f,-p,a,0,t) = {}_0d_t^{1-m}\psi_1(t) + {}_0d_t^{2-m}\psi_2(t) + \ldots {}_0d_t^{m-m}\psi_m(t) = \sum_{j=1}^{m}{}_0d_t^{j-m}\psi_j(t).$$

Then $L\left\{\frac{d^m}{dt^m}\psi(f,-p,a,0,t)\right\} = L\left\{\sum_{j=1}^{m}\frac{d^j\psi_j(t)}{dt^j}\right\} = \sum_{j=1}^{m}s^j L\{\psi_j(t)\}$. Here, redundant terms have been dropped. Therefore, (7.31) is further generalized as:

$$L\left\{{}_0D_t^u f(t)\right\} = L\left\{{}_0D_t^m {}_0D_t^{-p} f(t)\right\} = s^u L\{f(t)\} + \sum_{j=1}^{m}s^j L\{\psi_j(t)\}$$
$$+ L\{\psi(h,m,a,0,t)\},$$

the difference with (7.31) is that the order of 1 derivatives are each initialized by time varying functions. Meaning the order 1 differentiations is now generalized order 1 differentiation. The derivative initialization of (7.17) i.e. $L\{\psi(h,m,a,0,t)\}$ can also be similarly decomposed, to have similar effect.

7.4.2 Decomposition of Fractional Derivative in Fractional Order

Figure 7.5 demonstrates this concept as signal flow graph.

As the decomposition of fractional integration was done by breaking into several fractional elements $-q_k$ in the integration section, similarly by replacing all the fractional integration components $-q_k = r_k$, this process is carried out.

The result of Fig. 7.5 is

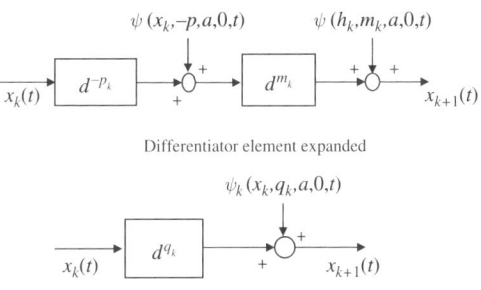

Fig. 7.5 Decomposition of fractional derivative (fractional differentiation element)

$$L\left\{{_0}D_t^q f(t)\right\} = L\left\{{_0}D_t^{B_1} x_1(t)\right\} = L\left\{x_{n+1}(t)\right\}$$

$$= s^{B_1} L\left\{x_1(t)\right\} + \psi_n(s) + \sum_{j=1}^{n-1} s^{B_{j+1}} \psi_j(s) B_a$$

$$= \sum_{i=a}^{n} r_i \tag{7.32}$$

$r_k \geq 0$ and $1 \leq a \leq n$, with initialization as:
$\psi_k(x_k, r_k, a, 0, t) = \frac{d^{m_k}}{dt^{m_k}} \psi(x_k, -p_{k-1}, a.0, t) + \psi(h_k, m_k, a, 0, t)$, where m_k is the least integer, such that $r_k = m_k - p_k$ and $h_k = {_0}d_t^{-p_k} x_k(t)$

7.4.3 Effect of Terminal Charging on Laplace Transforms

In all the above discussions, the initialization term ψ is taken as completely arbitrary, treating as "side charging" of the differintegrals. The effect of "terminal charging" is readily determined by appropriate substitutions for the ψ terms. That is for integer derivative make $\psi(h, m, a, 0, t) = 0, m = 1, 2\ldots$ and for the fractional integrations take $\psi(f, -p, a, 0, t) = \frac{1}{\Gamma(p)} \int_a^0 (t - \tau)^{p-1} f(\tau) d\tau$. Some modest simplifications are done to Laplace transform of the fractional integral, the term $\psi^{(k)}$ for terminal charging, as follows:

$$\psi^{(k)}(f, -q, a, 0, t)|_{t=0} = \left\{ \frac{d^{k-1}}{dt^{k-1}} \left(\frac{d}{dt} \int_a^0 \frac{(t-\tau)^{q-1}}{\Gamma(q)} f(\tau) d\tau \right) \right\}_{t=0}, \quad t > 0$$

Apply Lebniz's rule, and we have

$$\psi^{(k)}(f, -q, a, 0, t)|_{t=0} = \left\{ \frac{d^{k-2}}{dt^{k-2}} \left(\frac{d}{dt} \int_a^0 \frac{(q-1)(t-\tau)^{(q-1)-1}}{\Gamma(q)} f(\tau) d\tau \right) \right\}_{t=0}$$

Substitute $\frac{(q-1)}{\Gamma(q)} = \frac{1}{\Gamma(q-1)}$, then the integral is recognized as $(q-1)$th order differintegral and continuing the process gives

$$\psi^{(k)}(f, -q, a, 0, t)|_{t=0} = \psi(f, -q + k, a, 0, t)|_{t=0}$$

In a similar fashion, a more general term is to be substituted for terminal initial function as:

7.5 Start Point Shift Effect

$$\psi^{(n)}(f,-q,a,0,t) = \psi(f,-q+n,a,0,t) = \frac{1}{\Gamma(q-n)} \int_a^0 (t-\tau)^{q-n-1} f(\tau) d\tau$$

$t > 0$, and $(n+1 > q > n > 0)$

7.5 Start Point Shift Effect

7.5.1 Fractional Integral

Specifically in effect of start of function at non-zero a is considered, and differintegration start at c (non-zero) is shown, $c > a \geq 0$ and $q > 0$. Then Laplace transform of un-initialized fractional integral is

$$L\left\{_aD_t^{-q} f(t)\right\} = L\left\{_ad_t^{-q} f(t)\right\} = L\left\{\frac{1}{\Gamma(q)} \int_a^t (t-\tau)^{q-1} f(\tau) d\tau\right\}$$

By definition, $f(t) = 0$ for $t \leq a$, thus the following can be formulated by unit step at $t = a$ as $f(t) = u(t-a)f(t)$. The result is

$$L\left\{_aD_t^{-q} f(t)\right\} = L\left\{_ad_t^{-q} f(t)\right\} = L\left\{_0d_t^{-q}(u(t-a)f(t))\right\}$$

7.5.2 Fractional Derivative

Under the same condition as above, the un-initialized fractional derivative is considered, then $L\left\{_ad_t^q f(t)\right\} = L\left\{_ad_t^m{_ad_t^{-p}} f(t)\right\} = L\left\{\frac{d^m}{dt^m} \frac{1}{\Gamma(p)} \int_a^t (t-\tau)^{p-1} f(\tau) d\tau\right\}$, as before m is the least integer greater than q. Again by definition $f(t) = u(t-a)f(t)$. Thus

$$L\left\{_ad_t^q f(t)\right\} = L\left\{\frac{d^m}{dt^m} \frac{1}{\Gamma(p)} \int_0^t (t-\tau)^{p-1} u(\tau-a) f(\tau) d\tau\right\}$$

$$= L\left\{\frac{d^m}{dt^m} {_0d_t^{-p}}(u(t-a)f(t))\right\} = L\left\{_0d_t^q u(t-a)f(t)\right\}$$

So the final result is $L\left\{_ad_t^q f(t)\right\} = L\left\{_0d_t^q u(t-a)f(t)\right\}$.

7.6 Laplace Transform of Initialization Function

7.6.1 Fractional Integral

From shifting theorem of Laplace transform, we have

$$L\{g(t-a)u(t-a)\} = e^{-as}L\{g(t)\},$$

for $a > 0$. Taking $f(t) = g(t-a)$ we thus have,

$$L\{f(t)u(t-a)\} = e^{-as}L\{f(t+a)\}.$$

Now the fractional integral under terminal charging the desired Laplace transform will be

$$L\{\psi(f,-q,a,c,t)\} = L\left\{{}_ad_t^{-q}f(t)\right\} - L\left\{{}_cd_t^{-q}f(t)\right\}$$
$$= L\left\{{}_0d_t^{-q}f(t)u(t-a)\right\} - L\left\{{}_0d_t^{-q}f(t)u(t-c)\right\}$$
$$= s^{-q}L\{f(t)u(t-a\} - s^{-q}L\{f(t)u(t-c)\}$$

Finally, $L\{\psi(f,-q,a,c,t)\} = s^{-q}\left[e^{-as}L\{f(t+a)\} - e^{-cs}L\{f(t+c)\}\right].$

7.6.2 Fractional Derivative

The equivalent initialization function obtained for fractional derivative is

$$\psi_{eq}(f,u,a,c,t) = \frac{d^m}{dt^m}\psi(f,-p,a,c,t) + \psi(f,-p,a,c,t),$$

where m is the least integer greater than u. For terminal charging, it has been shown $\psi(h,m,a,c,t){=}0$, therefore $L\{\psi(f,u,a,c,t)\} = s^m L\{\psi(f,-p,a,c,t)\}$; in this, applying the shifting theorem of Laplace the final result is

$$L\{\psi(f,u,a,c,t)\} = s^u\left[e^{-as}L\{f(t+a)\} - e^{-cs}L\{f(t+c)\}\right].$$

7.7 Examples of Initialization in Fractional Differential Equations

Proper initialization is crucial in the solution and understanding of fractional differential equations; the examples will elucidate the application of initialized fractional calculus to the solution of differential equations.

7.7 Examples of Initialization in Fractional Differential Equations

Example 1.

$$_0D_t^{1/2} f(t) + bf(t) = 0 \quad t > 0; \quad _0D_t^{-1/2} f(t)_{at\ t=0} = C$$

The notation above is un-initialized semi-derivative and by obtaining Laplace transform will give $F(s) = \frac{C}{s^{1/2}+b}$. The inverse Laplace gives

$$f(t) = Ct^{-1/2} E_{0.5, 0.5}(-b\sqrt{t})$$

and for $b = 1$, $f(t) = C\left(\frac{1}{\sqrt{\pi t}} - e^t erfc\left(\sqrt{t}\right)\right)$.

Now the initialized approach is given below, considering semi- derivative as initialized.

$_0D_t^{1/2} f(t) + bf(t) = 0 \quad t > 0$. $\psi(f, 1/2, a, 0, t)$ is arbitrary, for 'side initialization' case. Therefore, the equation is written as

$$_0d_t^{1/2} f(t) + \psi(f, 1/2, a, 0, t) + bf(t) = 0, t > 0,$$

and $\psi(f, 1/2, a, 0, t)$ is arbitrary. Laplace transforms now gives

$$F(s) = \frac{-\psi(f, 1/2, a, 0, s)}{s^{1/2} + b} = -\frac{\psi(s)}{s^{1/2} + b}.$$

This Laplace transform is same as for un-initialized approach (done above) when $\psi(t) = -C\delta(t)$, i.e. "when impulse at $t = 0$ is used to initialize the fractional differential equation.' Now using R function (variant of Mittag-Leffler function), the generalized inverse of the initialized transform by applying

$$R_{q,v}(\alpha, c, t) \equiv \sum_{n=0}^{\infty} \frac{(\alpha)^n (t-c)^{(n+1)q-v-1}}{\Gamma((n+1)q - v)} \leftrightarrow \frac{s^v}{s^q - \alpha}$$

and with convolution integral definition, we obtain most general solution of the form as:

$$f(t) = -\int_0^t R_{1/2, 0}(-b, 0, t - \tau) \psi(\tau) d\tau \quad t > 0$$

Now with this arbitrary initialization, the above convolution integral is the most general solution. If $\psi(t) = -C\delta(t)$, from the derived general equation, we obtain $f(t) = CR_{1/2, 0}(-b, 0, t)$, which is identical to the above result of un- initialized case.

For 'terminal initialization,' a more useful result would appear using the terminal charging definition as:

$$\psi(f,q,a,0,t) = \frac{d^m}{dt^m}\psi(f,-p,a,0,t) + \psi(h,m,a,0,t)$$
$$q > 0 \quad h(t) = {}_ad_t^{-p}f(t)$$

In the present case $q = m - p$, $m = 1$, $p = 1/2$, and $\psi(h,m,a,0,t) = 0$. Then

$$\psi(f,1/2,a,0,t) = \frac{d}{dt}\psi(f,-1/2,a,0,t) = \frac{d}{dt}\frac{1}{\Gamma(1/2)}\int_a^0 (t-\tau)^{-1/2}f(\tau)d\tau$$

Example 2.

$$_0D_t^Q f(t) + {}_0D_t^q f(t) = h(t) \quad \left[{}_0D_t^{Q-1}f(t) + {}_0D_t^{q-1}f(t)\right]_{at\ t=0} = C$$

The problem uses symbolism as un-initialized fractional differential operator. Two separate initialization functions for the two fractional differential operators, when employed, gibes the most general solution as demonstrated below.

$$_0d_t^Q f(t) + \psi_1(f,Q,a,0,t) + {}_0d_t^q f(t) + \psi_2(f,q,a,0,t) = h(t)$$

and taking Laplace transform:

$$s^Q F(s) + s^q F(s) = H(s) - \psi_1(f,Q,a,0,s) - \psi_2(f,q,a,0,s)$$
$$F(s) = \left(\frac{s^{-q}}{s^{Q-q}+1}\right)[H(s) - \psi_1(f,Q,a,0,s) - \psi_2(f,q,a,0,s)]$$

again using generalized R function and the convolution definition, we get a most generalized form of solution:

$$f(t) = \int_0^t R_{Q-q,-q}(-1,0,t-\tau)(h(\tau) - \psi_1(f,Q,a,0,\tau) - \psi_2(f,q,a,0,\tau))d\tau.$$

In this, the general expression allows having effect of continuing past, as in terminal initialization. Giving these initialization function arbitrary values of say $\psi_1 = -C_1\delta(t)$.&. $\psi_2 = -C_2\delta(t)$ and putting $C = C_1 + C_2$, the general solution will be $f(t) = -CR_{Q-q,-q}(-1,0,t) + \int_0^t R_{Q-q,-q}(-1,0,t-\tau)(h(\tau))d\tau$.

It is to show that in this case of side charging, the effect of continuing past is not shown in the solution.

Example 3. Consider the same example as in Example 7.2, i.e.

$$_cD_t^Q f(t) + {}_cD_t^q f(t) = h(t)$$

This may be written as

7.8 Problem of Scalar Initialization

$$_c d_t^Q f(t) + _c d_t^q f(t) = h(t) - \psi_1(f, Q, a, c, t) - \psi_2(f, q, a, c, t) = h(t) - \psi_{eq}(t)$$

The Laplace transform of un-initialized fractional derivative is

$$L\left\{_c d_t^q f(t)\right\} = L\left\{_0 d_t^q (u(t-c) f(t)\right\} = e^{-cs} s^q L\left\{f(t+c)\right\},$$

where $u(t-c)$ is unit step function, at time $t = c$. Thus, the Laplace transform of the equation yields $L\{f(t+c)\} = \frac{H(s) - \psi_{eq}(s)}{e^{-cs} s^q (s^{Q-q}+1)} = e^{cs} G(s) \left[H(s) - \psi_{eq}(s)\right]$, where $G(s) = \frac{1}{s^q(s^{Q-q}+1)}$ Consider two time intervals. First interval $a_1 = c_1 = 0$ and second interval $a_2 = c_2 = 1$. The first interval is used as the initial period (charging) for the second interval. We take excitation (forcing function) as $h_1(t) = (u(t) - u(t-1))$, i.e. square window of height unity from 0 to 1. Also $f(t) = 0$, at $t < 0$, which also tells $\psi_1(f, Q, 0, 0, t) = \psi_2(f, q, 0, 0, t) = 0$.

First Interval In this interval, the solution is

$$L\{f_1(t)\} = \frac{1 - e^{-s}}{s^{q+1}(s^{Q-q} + 1)} = \left(\frac{1 - e^{-s}}{s}\right) G(s)$$

and $f_1(t) = R_{Q-q, -q-1}(-1, 0, t) - u(t-1) R_{Q-q, -q-1}(-1, 0, t-1) \quad t > 0$
for $0 < t < 1$ then $f_1(t) = R_{Q-q, -q-1}(-1, 0, t)$.

Second Interval In this interval, consider the forcing function $h(t) = 0$, and the interval 1 is the initialization interval. The equation for this interval is

$$_1 d_t^Q f_2(t) + _1 d_t^q f_2(t) = -\psi_{eq}(t).$$

Using Laplace we get

$$L\{f_2(t+1)\} = \frac{-\psi_{eq}(s)}{e^{-s} s^q (s^{Q-q} + 1)}$$

ψ_{eq} is taken based on the historic forcing $h_1(t)$ as was in the first interval. $\psi_{eq} = -h_1(t) = -(u(t) - u(t-1))$. Hence $\psi_{eq}(s) = -\frac{1-e^{-s}}{s}$. Substituting these, we get

$$e^{-s} L\{f_2(t+1)\} = \frac{1 - e^{-s}}{s^{q+1}(s^{Q-q} + 1)} = \frac{1 - e^{-s}}{s} G(s) = L\{f_1(t)\}$$

Applying the Laplace shift rule, we have $L\{f_2(t) u(t-1)\} = L\{f_1(t)\}$. Thus for $t > 1$, we have the important result $f_2(t) = f_1(t), t > 1$.

7.8 Problem of Scalar Initialization

Linear scalar constant coefficient fractional differential equation for $0 < q < 1$ is given as $_0 d_t^q x(t) = A x(t) + B u(t)$, assuming causality of the system, and the system was at rest meaning $x(t) = 0$ for all $t < a$. The qth derivative of $x(t)$ starts at time a and continues at time t. This means

$$_0d_t^q x(t) \equiv \frac{d}{dt}\left[\frac{1}{\Gamma(1-q)}\int_a^t \frac{x(\tau)}{(t-\tau)^q}d\tau\right].$$

Let the initialization of the scalar start at some time c after a. Choosing $c = 0$, the differential equation can be written as $_0d_t^q x(t) + {_ad_t^q}x(t) = Ax(t) + Bu(t)$. Here the time axis is broken in two parts, one from $a \to 0$ and the other from $0 \to t$. Here as stated $c = 0$, is chosen. The term $_ad_0^q x(t)$ above represents this initialization response due to behavior of system before $t = c = 0$. It should be noticed that the past history of the particular variable that is fractionally differentiated must be known as long as the system has been operated to obtain the correct initialization response. The equation then can be expressed as in terms of initialized fractional derivative as: $_cD_t^q x(t) = {_0d_t^q}x(t) + \psi(q, x, a, 0, t) = Ax(t) + Bu(t)$, where for terminal charging for $t > 0$, the initialization function is

$$\psi(q, x, a, 0, t) \equiv {_ad_0^q}x(t) = \frac{d}{dt}\left[\frac{1}{\Gamma(1-q)}\int_a^0 \frac{x(\tau)}{(t-\tau)^q}d\tau\right].$$

This initialization function is described in Chap. 6, called the initialization function for fractional derivative. Using the Laplace transforms of the initialized expression, we obtain the following:
$s^q X(s) + \psi(q, x, a, 0, s) = AX(s) + BU(s)$, where

$$\psi(q, x, a, 0, s) = L\left\{\frac{d}{dt}\left[\frac{1}{\Gamma(1-q)}\int_a^0 \frac{x(\tau)}{(t-\tau)^q}d\tau\right]\right\}$$

is the Laplace transform of the initialized function. Rearranging the above expression, we get

$$X(s) = \frac{B}{s^q - A}U(s) - \frac{1}{s^q - A}\psi(q, x, a, 0, s).$$

Taking inverse Laplace transform of this expression, we obtain time response:
$x(t) = \int_0^t F_q[A, \tau]Bu(t-\tau)d\tau - \int_0^t F_q[A, \tau]\psi(q, x, a, 0, t-\tau)d\tau$. Here F function is the impulse response of fundamental linear differential equation (as explained in Chap. 6), and is defined as

$$F_q[A, t] \equiv t^{q-1}\sum_{n=0}^{\infty}\frac{A^n t^{nq}}{\Gamma(nq+n)}.$$

The solution $x(t)$ represents any forced response due to $u(t)$, and the second term of the solution expresses the initialized response of the system due to past history of $x(t)$, before time $t = 0$. Clearly for integer order systems ($q = 1$), this initialization term $\psi(1, x, a, 0, s)$ equals a constant and for fractional order systems, this term $\psi(1, x, a, 0, s)$ is a time varying expression into the future. It implies that the past history of $x(t)$ has appearance of a time-dependent forcing term into the "infinite" future.

Let us choose the history as a constant meaning $x(t) = k$, for $-\infty < t \leq 0$. Then the initialization function becomes the limit $a \to \infty$, described as follows:

$$\psi(q, x, -\infty, 0, s) = L\left\{\lim_{a \to -\infty} \frac{d}{dt}\left[\frac{1}{\Gamma(1-q)} \int_a^0 \frac{k}{(t-\tau)^q} d\tau\right]\right\}$$

$$= L\left\{\lim_{a \to -\infty} \frac{d}{dt}\left[\frac{k}{\Gamma(1-q)} \frac{-(t-\tau)^{1-q}}{(1-q)}\right]_{\tau=a}^{\tau=0}\right\}$$

$$= L\left\{\lim_{a \to -\infty} \frac{d}{dt}\left[\frac{k}{\Gamma(1-q)}\left(\frac{(t-a)^{1-q}}{(1-q)} - \frac{t^{1-q}}{(1-q)}\right)\right]\right\}$$

$$= L\left\{\lim_{a \to -\infty} \frac{k}{\Gamma(1-q)}\left(\frac{1}{(t-a)^q} - \frac{1}{t^q}\right)\right\}$$

$$= -L\left\{\frac{k}{\Gamma(1-q)} \frac{1}{t^q}\right\} = -ks^{q-1}$$

Inserting this back into Laplace expression, we obtain $X(s) = \frac{B}{s^q - A} U(s) + \frac{ks^q}{s(s^q - A)}$. Inverting this, we obtain the solution:

$$x(t) = B \int_0^t F_q[A, q] u(t-\tau) d\tau + k E_q\left[At^q\right].$$

The first term is convolution of the input $u(t)$ with impulse response (F function), and the second term is the initialization function response, and in this particular case of history (k) is Mittag-Leffler function. For integer order systems $q = 1$ the initialized Laplace term (second term of $X(s)$) in case of the constant is k/s.

7.9 Problem of Vector Initialization

Vector space representation is useful for "systems of fractional differential equations." Once the minimal basis value q is chosen the vector representation can be expressed as: ${}_cD_t^q \bar{x}(t) = A\bar{x}(t) + B\bar{u}(t)$, the vector $\bar{x}(t)$ is given for $a \leq t \leq c$ or initialization vector $\bar{\psi}(q, \bar{x}, a, c, t)$ given for $t > c$, and the output vector is $\bar{y}(t) = C\bar{x}(t) + D\bar{u}(t)$. The parameters (matrices) A, B, C, D are usual state-space representation for systems of differential equations as for integer order systems, representing state matrix, input and output matrix, and feed-through matrix. The vector fractional order differential equation expressed above can be written as: $_c d_t^q \bar{x}(t) + \bar{\psi}(q, \bar{x}, a, 0, t) = A \bar{x}(t) + B \bar{u}(t)$, where

$$\bar{\psi}(q, \bar{x}, a, 0, t) = [\psi(q, x_1, a, 0, t), \psi(q, x_2, a, 0, t) \ldots \psi(q, x_n, a, 0, t)]^T$$
$$= [\psi_1, \psi_2, \ldots, \psi_n]^T.$$

At this point, it is important to notice that the fractional dynamic variable in the system of vector space equations are not states in true sense of the name state-space control terminology. In usual integer order system theory, the set of state of the system, known at any given point in time along with the system equations, are sufficient to predict the response of the system both forward or backward in time. The collection of constant numbers $\bar{x}(t_0)$ at time t_0 specifies the complete "state" of the system at that time. Therefore, the system will have unique time response, given its "initial-state."

Fractional dynamic variables do not represent the state of system at any given time (alone) due to presence of the initialization function vector (history function), carrying information about the history of the elements of the system. Consequently, as the initialization function vector is generally present, the set of elements of the vector $\bar{x}(t)$, evaluated at any point in time, does not specify the entire "state" of the system. Thus in fractional system setting, the ability to predict the future response of a system requires the set of fractional differential equations along with the initialization function sets. The initialization problem of the vector fractional differential equations can be solved as solved for the scalar case (Sect. 7.8), by using Laplace transformation.

The Laplace transformed vector equation is $s^q \bar{X}(s) + \bar{\psi}(s) = A \bar{X}(s) + B \bar{U}(s)$ and the output equation is transformed as $\bar{Y}(s) = C \bar{X}(s) + D \bar{U}(s)$. After performing matrix algebraic manipulations, we obtain $(I s^q - A) \bar{X}(s) = B \bar{U}(s) - \bar{\psi}(s)$, where I is $n \times n$ identity matrix matching dimensions of state matrix A. Therefore, the Laplace vector solution is thus obtained as

$$\bar{X}(s) = \left(I s^q - A\right)^{-1} B \bar{U}(s) - \left(I s^q - A\right)^{-1} \bar{\psi}(s),$$

inserting this in output expression to have output Laplace solution as

$$\bar{Y}(s) = \left\{ C \left(I s^q - A\right)^{-1} B + D \right\} \bar{U}(s) - C \left(I s^q - A\right)^{-1} \bar{\psi}(s).$$

Inverting this one, we obtain time response as

$$\bar{y}(t) = C \int_0^t F_q [A, \tau] \left\{ B \bar{u}(t - \tau) - \bar{\psi}(t - \tau) \right\} d\tau + D \bar{u}(t)$$

or equivalently can be expressed as

$$\bar{y}(t) = C \int_0^t F_q[A, t-\tau]\left\{B\bar{u}(\tau) - \bar{\psi}(\tau)\right\} d\tau + D\bar{u}(t).$$

The above solution requires the use of matrix F function, which can be obtained by the use of its series expansion. Matrix F function is defined as

$$F_q[A, t] \equiv t^{q-1} \sum_{n=0}^{\infty} \frac{A^n t^{nq}}{\Gamma(nq+n)}$$

for $q > 0$,
where A is $n \times n$ system matrix.
Consider the Laplace transformed equation

$$\bar{Y}(s) = \left\{C(Is^q - A)^{-1}B + D\right\}\bar{U}(s) - C(Is^q - A)^{-1}\bar{\psi}(s),$$

derived earlier in this section. In this case, the system transfer matrix is $\bar{G}(s) = \left\{C(Is^q - A)^{-1}B + D\right\}$, and is system representation of multivariate system. In this representation thus the system description is obtained as $\bar{Y}(s) = \bar{G}(s)\bar{U}(s) - C(Is^q - A)^{-1}\bar{\psi}(s)$. The A, B, C, D are usual state-space representation matrix namely system state matrix, input matrix, output matrix, and feed-through matrix, as per multivariate state-vector representation of modern control science.

7.10 Laplace Transform $s \to w$ Plane for Fractional Controls Stability

As it is difficult to visualize multiple Riemann sheets, it is useful to perform "conformal transformation" into another complex plane $s \to w$ plane. System dynamics are described by singularity (pole) location of the transformed transfer function in new w-plane. The transfer function $G(s) = \frac{1}{s^q+1}$, $(0 < q < 1)$ does not have any singularity (pole) in anywhere in s-plane (primary Riemann sheet), but after crossing branch-cut the secondary Riemann sheet will contain the singularity (pole). For $q = 1/2$, the denominator $(s^{1/2} + 1)$ goes to zero at $s = 1 + j0 = \exp(\pm j2\pi)$, which is underneath the s-plane negative real axis $\exp(\pm j\pi)$, entering "secondary Riemann sheet", giving pole of the function there.

Here in this section, basic control theory knowledge is required, in Laplace domain. To understand the behavior and stability property of the fundamental fractional differential equation $_0d_t^q x(t) = -ax(t) + bu(t)$, it is necessary to analyze the pole-location of the system transfer function i.e. $G(s) = \frac{X(s)}{U(s)} = \frac{b}{s^q+a}$. For classical control theory $q = 1$, that is integer order system, the pole-location are studied in the complex Laplace plane (s-plane). The stability boundary in the s-plane is the imaginary axis. Any pole lying to the right of the imaginary axis represents an unstable time response. Examining $G(s) = \frac{b}{s^q+a}$, however, indicates that the poles of $G(s)$ must now be evaluated in s^q plane. Rather than dealing with fractional powers of s, the analysis is carried out in $w = s^q$, and then the pole-location properties will be studied in the new complex w-plane.

To simplify the discussion, we limit the fractional order as $0 < q \leq 1$. The mapping from $s \to w$ is as follows:

$$w = \rho e^{j\phi} = \alpha + j\beta \text{ and } s = re^{j\theta}.$$

Then defining $w = s^q = \left(re^{j\theta}\right)^q = r^q e^{jq\theta} = \rho e^{j\phi}$, implying $\rho = r^q$ and $\phi = q\theta$.

With this equation, it is possible to map either lines of constant radius or lines of constant angle from the s-plane into w-plane. Of particular interest is the image of the s-plane stability boundary (s-plane imaginary axis), that is $s = re^{\pm j\frac{\pi}{2}}$, maps as $w = r^q e^{\pm jq\frac{\pi}{2}}$. This is a pair of lines at $\phi = \pm q\pi/2$. Thus, the right half plane (RHP) of the s-plane maps into a "wedge" in the w-plane of angle less than $\pm 90q$ degree that is RHP of s-plane maps into $|\phi| < q\pi/2$.

As an example for semi-differential equation of order $1/2$, the RHP of s-plane maps into the wedge bounded by $|\phi| < \pi/4$. A half-order system with its w-plane poles in the wedge that is $|\phi| < \pi/4$ would be unstable and corresponding F function (impulse response) would grow without bound.

It is also important to consider the mapping of the negative real axis of s-plane $s = re^{\pm j\pi}$, the mapping is $w = r^q e^{\pm jq\pi}$. Therefore, the entire "primary sheet" of the s-plane maps into a w-plane wedge of angle less than $\pm 180q$ degree; while all

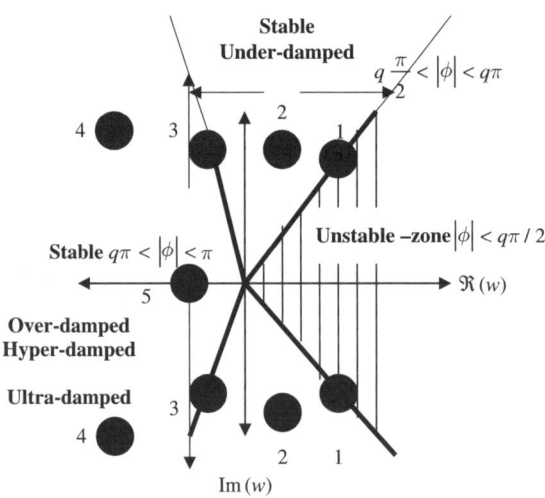

Fig. 7.6 w-Plane stability zones and various pole locations

Table 7.1 Properties of poles in w-plane

pole number (Fig. 7.6)	region in w-plane	region in s-plane	property of time-response
1	$\|\phi\| < q\frac{\pi}{2}$	$\Re e(s) > 0$ (RHP)	unstable
2	$q\frac{\pi}{2} < \|\phi\| < q\pi$	$\Re e(s) < 0$ (LHP)	stable under-damped oscillatory
3	$\phi = q\pi$	negative real axis $s = r \exp(\pm j\pi)$	stable over-damped
4	$q\pi < \|\phi\| < \pi$	secondary Riemann sheet $\|\theta\| > \pi$	stable hyper-damped
5	$\|\phi\| = \pi$	Secondary Reiman sheet	stable ultra-damped

the "secondary sheet" s-plane maps into the remainder of the w-plane. For half-order system, the negative real axis of s-plane maps into w-plane lines at ± 90 degree. Figure 7.6 gives the pole location properties of w-plane.

F function time response $F_q[a, t]$ depends on both q and on parameter a, which is the pole location of the system transfer function $G(s) = \frac{b}{s^q + a}$. For a fixed value of q, the angle ϕ of the parameter a, as measured from the positive real axis of w-plane, determines the type of response to expect. This pole location is depicted in the Table 7.1 with properties of time response.

All usual control system analysis Nyquist, root-locus plot concerning poles or eigenvalues can be used in the w-plane, remembering the stability boundary is now the image s-plane imaginary axis. Hyper-damped and ultra-damped system will be taken up again in Chap. 10.

7.11 Rational Approximations of Fractional Laplace Operator

There are lots of approaches to get rational approximation of fractional order Laplace operator. The s^q keeps on appearing for fractional order systems and controls. The rational function approximation gives direction to realize this fractional order Laplace operator in circuit impedance and admittance forms. This section gives insight into simple method of approximating the fractional Laplace operator. It is well known that for interpolation or evaluation purposes, rational functions are (sometimes) superior to polynomial fit because of their ability to model the function with poles and zeros. In other words, for evaluation purposes, rational approximations frequently converge much more rapidly than power series expansion (PSE) and have wider domain of convergence in the complex plane. These approximations can be viewed in the Laplace domain, as rational approximations of the fractional order operators. Furthermore, these approximations exhibit a common feature, which we observe in all good rational approximations: They have poles and zeros interlaced on the negative real axis of the s-plane, and the distance between successive poles and zeros decrease as the approximations are improved by increasing the degree of the numerator and denominator polynomial.

There are a number of methods to evaluate rational approximations of the fractional Laplace operator; they are general continued fractional expansion (CFE) method, Carlson's method, Matsuba's method, Oustaloup's method, Chareff's method, S.C. Dutta Roy's method, Wang's method, and Jones method.

Here, the Carlson's approach using Newton's approximation to find arbitrary root of polynomial is described. The method is derived from a regular Newton process for iterative approximation of the qth root. The starting point of the method is $(H(s))^{1/q} - (G(s)) = 0$, meaning $H(s) = (G(s))^q$. Defining $q = 1/p$ and $m = p/2$, the iteration formula starting from first approximation $H_0(s) = 1$ is

$$H_i(s) = H_{i-1}(s) \frac{(p-m)(H_{i-1}(s))^2 + (p+m)G(s)}{(p+m)(H_{i-1}(s))^2 + (p-m)G(s)}$$

Say we want to approximate semi-integral Laplace operator $\left(\frac{1}{s}\right)^{1/2}$ to be realized by this method, then take $G(s) = \frac{1}{s}$, $q = \frac{1}{2}$, it makes $p = 2$ and $m = 1$. We get $H_1(s)$ as the first rational approximate for initial function $H_0(s) = 1$ as:

$$H_1(s) = H_0(s) \frac{(p-m)(H_0(s))^2 + (p+m)G(s)}{(p+m)(H_0(s))^2 + (p-m)G(s)} = (1) \frac{(1)(1) + (3)\left(\frac{1}{s}\right)}{(3)(1) + (1)\left(\frac{1}{s}\right)} = \frac{s+3}{3s+1}$$

The second approximate is obtained by putting the obtained $H_1(s)$ as:

$$H_2(s) = \left(\frac{s+3}{3S+1}\right) \left[\frac{(1)\left(\frac{s+3}{3s+1}\right)^2 + (3)\left(\frac{1}{s}\right)}{(3)\left(\frac{s+3}{3s+1}\right)^2 + (1)\left(\frac{1}{s}\right)} \right] = \frac{(s+3)(s^3 + 33s^2 + 27s + 3)}{(3s+1)(3s^2 + 27s^2 + 33s + 1)}$$

The second approximate for semi-integration Laplace operator is

$$H_2(s) = \frac{s^4 + 36s^3 + 126s^2 + 84s + 9}{3s^4 + 84s^3 + 126s^2 + 36s + 1} \approx \left(\frac{1}{s}\right)^{1/2}.$$

Physical of this approximation steps is to be viewed as impedance (rather immitance), realized by R, L, C network. The first approximation $H_0(s)$ is unit impedance of pure resistance (unit value). The next approximation is $H_1(s)$, is RC combination, and as the approximation grows, they get more and more components, in the network. The fundamentals of circuit synthesis to realize these fractional Laplace operators in network forms and present day Hybrid Micro Circuit fabrication techniques will give electrical circuit components for fractional order circuit applications, in future.

Table 7.2 gives rational approximates for several fractional order Laplace operator for fractional integration realization.

7.12 Concluding Comments

Table 7.2 Fractional operators with approximately 2 db error from 0.01 to 100 rad/s

Laplace fractional operator	rational approximate
$\dfrac{1}{s^{0.1}}$	$\dfrac{220.4s^4 + 5004s^3 + 5038s^2 234.5s + 0.4840}{s^5 + 359.8s^4 + 5742s^3 + 4247s^2 147.7s + 0.2099}$
$\dfrac{1}{s^{0.2}}$	$\dfrac{60.95s^4 + 816.9s^3 + 582.8s^2 + 23.24s + 0.04934}{s^5 + 134.0s^4 + 956.5s^3 + 383.5s^2 + 8.953s + 0.01821}$
$\dfrac{1}{s^{0.3}}$	$\dfrac{23.76s^4 + 224.9s^3 + 129.1s^2 4.733s + 0.01052}{s^5 + 64.51s^4 + 252.2s^3 + 63.61s^2 + 1.104s + 0.002276}$
$\dfrac{1}{s^{0.4}}$	$\dfrac{25.00s^4 + 558.5s^3 + 664.2s^2 + 44.15s + 0.1562}{s^5 + 125.6s^4 + 840.6s^3 + 317.2s^2 + 7.428s + 0.02343}$
$\dfrac{1}{s^{0.5}}$	$\dfrac{15.97s^4 + 593.2s^3 + 1080s^2 + 135.4s + 1}{s^5 + 134.3s^4 + 1072s^3 + 543.4s^2 + 20.10s + 0.1259}$
$\dfrac{1}{s^{0.6}}$	$\dfrac{8.579s^4 + 255.6s^3 + 405.3s^2 35.93s + 0.1696}{s^5 + 94.22s^4 + 472.9s^3 + 134.8s^2 + 2.639s + 0.009882}$
$\dfrac{1}{s^{0.7}}$	$\dfrac{5.406s^4 + 177.6s^3 + 209.6s^2 + 9.197s + 0.01450}{s^5 + 88.12s^4 279.2s^3 + 33.30s^2 + 1.927s + 0.0002276}$
$\dfrac{1}{s^{0.8}}$	$\dfrac{5.235s^3 + 1453s^2 + 5306s + 254.9}{s^4 + 658.1s^3 + 5700s^2 + 658.2s + 1}$
$\dfrac{1}{s^{0.9}}$	$\dfrac{1.766s^2 + 38.27s + 4.914}{s^3 + 36.15s^2 + 7.789s + 0.01000}$

These approximate representations are utilized to have impedance structures for realization of fractional capacitance and with thus operational amplifier circuit technique, the fractional order integrator and fractional order differentiator be realized.

7.12 Concluding Comments

Many of the forms described in this chapter on the Laplace transform have been derived to demonstrate how initialization function generalizes the Laplace transforms (seen in integer order calculus). Here also several possible compositions of fractional integration and fractional derivatives are demonstrated. Reader is advised to imagine further, regarding the compositions of these operators, as to adhere to constraints of continuity and index (composition) laws to use these forms. The selection of appropriate initialization function when structuring fractional differential equations will be analogous, but somewhat more demanding, than selection of constants when structuring integer order systems. Experience working with particular equation or type of physical systems will be required of analyst in selection of initialization function. The use of Laplace transforms and initialization is obtained for scalar and vector initialization examples, it will be suitable for fractional order system modeling, for single input single output, as well as multi-input multi-output systems. The analysis of stability of the pole (and zero) property is to be carried out in transformed Laplace (w) plane, and the theory and interpretation of control system be mapped here, for future use of fractional order control system. Also the Laplace operator for fractional capacitor is represented as rational function approximation to synthesize circuits for fractional order control system, in analog way. These concepts are very useful for futuristic control systems.

Chapter 8
Application of Generalized Fractional Calculus in Electrical Circuit Analysis

8.1 Introduction

The fractional calculus is widely popular, especially in the field of viscoelasticity. In this chapter, a variety of applications are discussed. This chapter is application oriented to demonstrate the fundamentals of generalized (fractional) calculus developed earlier, with particular reference to initialization concepts. Here the treatment is to show coupling effect of the initialization functions and the use of developed Laplace technique. The applications and potential applications are in diffusion process, electrical science, electrochemistry, material creep, viscoelasticity, control science, electromagnetic theory, etc. This chapter is restricted to electronics and electrical circuit models.

8.2 Electronics Operational Amplifier Circuits

8.2.1 Operational Amplifier Circuit with Lumped Components

The operational amplifier is well known in electrical science to provide gain. Figure 8.1 gives a simple scheme.

We revise the concept of impedance transfer function. The terms Z_f and Z_i are output and input impedances respectively, will represent general linear circuit together with initialization functions. Therefore not strictly as impedance but some similarity to perhaps memory element!

The Laplace transforms variables:

$$\text{Input side: } v_i(s) - v_-(s) = i_i(s)Z_i(s)$$
$$\text{Output/feedback element: } v_-(s) - v_o(s) = i_f(s)Z_f(s)$$
$$\text{Inverting node: } i_i(s) - i_f(s) - i_-(s) = 0$$

Fig. 8.1 Operational amplifier circuit with lumped impedances

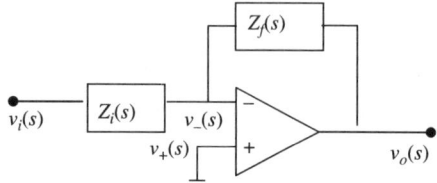

For negative feedback configuration, the amplifier will have from its basic property of high input impedance of NI (+) and I (−) terminals $i_- \cong 0$ therefore $v_- \cong 0$. Putting these values, we obtain transfer function (voltage to voltage) as:

$$\frac{v_o(s)}{v_i(s)} = -\frac{Z_f(s)}{Z_i(s)}$$

8.2.2 Operational Amplifier Integrator with Lumped Element

Figure 8.2 describes the classical integrator circuit.
The defining equations are

$$v_i(t) - v_-(t) = i_i R$$

$$v_-(t) - v_o(t) = \frac{1}{C} {}_cD_t^{-1} i_f(t) = \frac{1}{C}\int_c^t i_f(\tau)d\tau + \frac{1}{C}\int_a^c i_f(\tau)d\tau$$

$$= \frac{1}{C}\int_c^t i_f(\tau)d\tau \frac{1}{C}q(c) = \frac{1}{C}\int_c^t i_f(\tau)d\tau + \left[v_-(c) - v_o(c)\right]$$

Substitution of initial charge stored from time $t = a$ to time $t = c$ as integral of current is done. $q(c) = \int_a^c i_f(\tau)d\tau = C\left[v_-(c) - v_o(c)\right]$ as initial charge on the capacitor at the start of integration process at time $t = c$. For operational amplifier, putting $i_f = i_i$ and $v_-(t) = 0$ yields

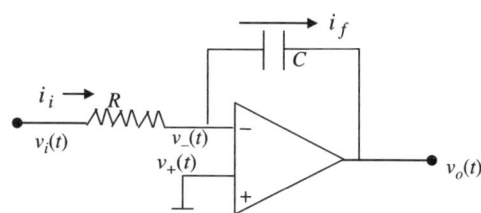

Fig. 8.2 Integrator circuit with lumped element

8.2 Electronics Operational Amplifier Circuits

$$v_o(t) = -\frac{1}{RC}\int_c^t v_i(\tau)d\tau + v_o(c) = -\frac{1}{RC}{}_cD_t^{-1}v_i(t)$$

with $\psi(v_i, -1, a, c, t) = -RCv_o(c)$ as initializing function.

Then with $RC = 1$, this integrator is represented as $v_o(t) = -{}_cD_t^{-1}v_i(t)$. The initialization function as defined stores the past history (contained in single constant $-v_o(c)$).

When this circuit is used in analog PID controller, the integral reset rate is defined as $T_i = \left|\frac{Z_f}{Z_i}\right| = \frac{1}{RC}$ in unit of per second.

8.2.3 Operational Amplifier Integrator with Distributed Element

The input resistor is replaced by semi-infinite lossless transmission line, with distributed inductance per unit length l and capacitance per unit length c. The semiinfinite lossless line has a characteristics equation as the wave equation of voltage, as:

$$\frac{\partial^2 v(x,t)}{\partial t^2} = \frac{1}{lc}\frac{\partial^2 v(x,t)}{\partial x^2}$$

The terminal characteristics is

$$i(t) = \sqrt{\frac{c}{l}}v(t) + \varphi_1(t) = kv(t) + \varphi_1(t), \text{ where } k = \sqrt{c/l}$$

or

$$i(t) = k_c D_t^0 v(t), \text{ where } \psi(v, 0, a, c, t) = \frac{1}{k}\varphi_1(t)$$

or

$$v(t) = \frac{1}{k}i(t) + \varphi_2(t)$$

or

$$v(t) = \frac{1}{k}{}_cD_t^0 i(t) \text{ where } \psi(i, 0, a, c, t) = k\varphi_2(t)$$

Notice that differentigration operation for voltage–current transfer relationship is of order zero. This may be called a "zero-order" element. The voltage–current relationship of a lumped resistor is also of zero-order differintegration. In the lossless transmission line, it is the initial condition on the distributed inductor and capacitor along infinite lines that gives rise to initial function (of time). Figure 8.3

Fig. 8.3 Semi-infinite lossless line (zero-order differintegrator)

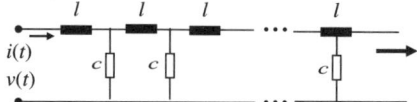

gives the semi-infinite lossless (zero-order) element representation and Fig. 8.4 is represents the operational amplifier circuit based on this lossless line.

The semi-infinite lossless line is connected as input element between the input terminal and inverting node. The input element equation is

$$i_i(t) = k\left[v_i(t) - v_-(t)\right] + \varphi_1(t)$$

For feed back element

$$v_-(t) - v_o(t) = \frac{1}{C}{}_cD_t^{-1}i_f(t) = \frac{1}{C}\int_c^t i_f(\tau)d\tau - v_o(c)$$

Putting $i_f = i_i$ and $v_-(t) = 0$, we have expression for output as:

$$v_o(t) = -\frac{1}{C}\int_c^t [kv_i(\tau) + \varphi_1(\tau)]d\tau + v_o(c)$$

$$= -\left(\frac{1}{C}k\right)\int_c^t v_i(\tau)d\tau - \frac{1}{C}\int_c^t \varphi_1(\tau)d\tau + v_o(c)$$

In compact generalized calculus notation, we summarize the transfer expression as:

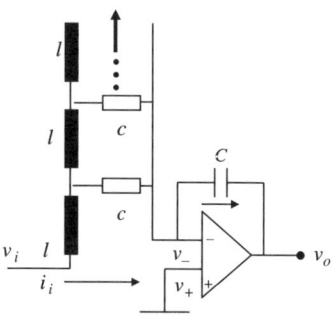

Fig. 8.4 Integrator circuit with distributed elements (one-order integrator)

8.2 Electronics Operational Amplifier Circuits

$$v_o(t) = -\left(\frac{1}{C}k\right){}_cD_t^{-1}v_i(t) \text{ where } \psi(v_i, -1, a, c, t) = \frac{1}{k}\int_c^t \varphi_1(\tau)d\tau + C\frac{1}{k}v_o(c)$$

Selecting the coefficient values and circuit constants as unity we have $v_o(t) = -{}_cD_t^{-1}v_i(t)$. This is the same expression that is got in classical integrator circuit with lumped elements. The difference exists in the values of the initialization functions. For this integrator with distributed elements, the effect of "past history" is contained not in a constant $-v_o(c)$, which is the charge stored in the integrating capacitor C (the lumped element), but is in the remainder of the function $\psi(t)$. This accounts for the distributed charge along the semi-infinite line. The observation is that the "zero-order differintegral" element that is the semi-infinite lossless line will simply propagate the changes in input excitation ($v_i(t)$) along the infinite line, and never this perturbation will be seen again. The only effect seen being proportional to the variation of the source end current ($i_i(t)$). This is true for "terminal charging." For side charging or a "side charged" line, an additional function of time may be impressed at output point of the circuit, which is dependent on the initial voltage distribution on the line.

8.2.4 Operational Amplifier Differential Circuit with Lumped Elements

Figure 8.5 describes the circuit, which is the classical differentiator block.
The expressions are for this classical differentiator:

$$v_i(t) - v_-(t) = \frac{1}{C}{}_cD_t^{-1}i_i(t) = \frac{1}{C}\int_c^t i_i(\tau)d\tau + \frac{1}{C}\int_a^c i_i(\tau)d\tau$$

$$= \frac{1}{C}\int_c^t i_i(\tau)d\tau + \frac{q(c)}{C} = \frac{1}{C}\int_c^t i_i(\tau)d\tau + v_{i-(-)}(c)$$

$$v_-(t) - v_o(t) = i_f(t)R$$

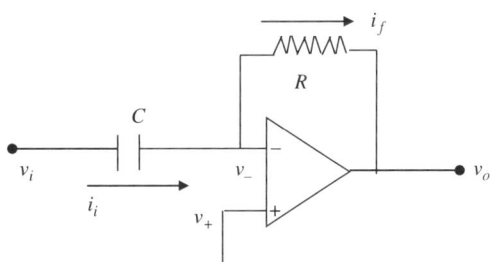

Fig. 8.5 Differentiator circuit with lumped elements

Putting $i_i(t) = i_f(t) = 0$ and $v_-(t) = 0$, we have

$$v_o(t) = -Ri_f(t) = -RC\left[\frac{d}{dt}(v_i(t) - v_{i-(-)}(c))\right]$$

In generalized calculus terms for $t > c$, we get $v_o(t) = -RC\,_cD_t^1 v_i(t)$. The initialization function to this generalized derivative is $\psi(v_i, 1, a, 0, t) = \frac{d}{dt}v_{i-(-)}(c)$. This initialization is normally put as zero. However, this initialization term can give a "pulse" response at time $t = c$.

8.2.5 Operational Amplifier Differentiator with Distributed Element

Figure 8.6 is the circuit where at the feedback element of the circuit is replaced by semi-infinite lossless transmission line (zero-order distributed element).

The following is the derivation for the differentiator with distributed zero-order elements:

$$v_i(t) - v_-(t) = \frac{1}{C}\,_cD_t^{-1}i_i(t) = \frac{1}{C}\int_c^t i_i(\tau)d\tau + v_{i-(-)}(c)$$

The distributed lossless (zero-order) element is connected at feedback path and using its relation we get

$$v_-(t) - v_o(t) = \frac{1}{k}\,_cD_t^0 i_f(t) = \frac{1}{k}i_f(t) + \frac{1}{k}\psi(i_f, 0, a, c, t), \text{ with } k = \sqrt{\frac{c}{l}}$$

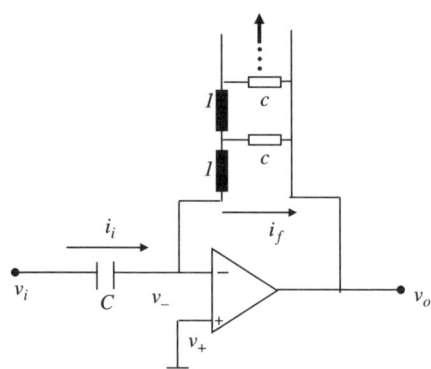

Fig. 8.6 Differentiator circuit with distributed element (one-order differentiator)

Putting $i_i(t) = i_f(t)$ and $v_-(t) = 0$, we have the differentiator expression:

$$v_o(t) = -\frac{1}{k}\left[C\frac{d}{dt}\left(v_i(t) - v_{i-(-)}(c)\right) + \psi(i_f, 0, a, c, t)\right] \text{ or}$$

$$v_o(t) = -C\frac{1}{k}\left[\frac{d}{dt}v_i(t) + \frac{1}{C}\psi(v_i, 1, a, c, t)\right]$$

In terms of generalized calculus notations and concepts, the differentiator circuit is described by $v_o(t) = -C\frac{1}{k}{}_cD_t^1 v_i(t)$, where initialization function associated with the derivative operator is $\psi(v_i, 1, a, c, t) = -\frac{1}{C}\psi(i_f, 0, a, c, t)$. As in the integrator case with distributed zero-order elements, the perturbations at the source terminal end will be propagated away and never return back (reflected). It is noted that for terminally charged integer order differentiator with lumped elements $\psi_{eff} = 0$.

8.2.6 Operational Amplifier as Zero-Order Gain with Lumped Components

In Fig. 8.1 consider lumped impedances as pure lumped resistances. Therefore, the circuit transfer function is
$v_o(t) = -\frac{R_f}{R_i}v_i(t) = -\frac{R_f}{R_i}{}_cD_t^0 v_i(t)$, where clearly $\psi(v_i, 0, a, c, t) = 0$. Meaning that this circuit configuration has no memory. Also it is a classical zero-order operator.

8.2.7 Operational Amplifier as Zero-Order Gain with Distributed Elements

Figure 8.7 gives the circuit diagram where feedback element and the input element of Fig. 8.1 are constituted by distributed element as semi-infinite lossless line.

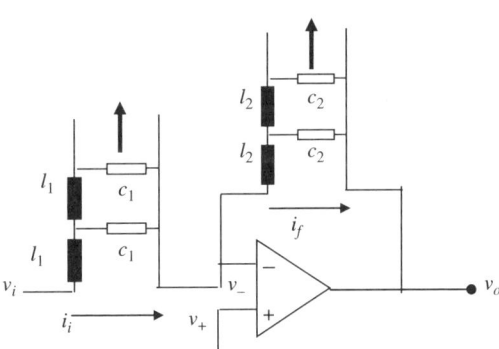

Fig. 8.7 Gain circuit with distributed elements (zero-order differintegration)

The describing relations are

$$i_i(t) = \sqrt{\frac{c_1}{l_1}} \,_cD_t^0 \left[v_i(t) - v_-(t) \right],$$

where initialization $\psi(v_i - v_-, 0, a, c, t) = \sqrt{\frac{l_1}{c_1}} \varphi_1(t)$.

$$v_-(t) - v_o(t) = \sqrt{\frac{l_2}{c_2}} \,_cD_t^0 i_f(t),$$

where the initialization $\psi(i_f, 0, a, c, t) = \sqrt{\frac{c_2}{l_2}} \varphi_2(t)$.

Putting $i_i = i_f$ and $v_-(t) = 0$, we have the expression:

$$v_o(t) = -\sqrt{\frac{l_2}{c_2}} \,_cD_t^0 \left[\sqrt{\frac{c_1}{l_1}} \,_cD_t^0 v_i(t) \right] = -\sqrt{\frac{c_1 l_2}{c_2 l_1}} \,_cD_t^0 v_i(t)$$

$$= -\sqrt{\frac{l_2}{c_2}} \,_cd_t^0 \left[\sqrt{\frac{c_1}{l_1}} \left\{ _cd_t^0 v_i(t) + \psi(v_i - v_-, 0, a, c, t) + \psi(i_f, 0, a, c, t) \right\} \right]$$

$$= -\sqrt{\frac{c_1 l_2}{c_2 l_1}} \,_cd_t^0 v_i(t) - \sqrt{\frac{l_2}{c_2}} \varphi_1(t) - \varphi_2(t)$$

Summarily, $v_o(t) = -\sqrt{\frac{c_1 l_2}{c_2 l_1}} \,_cD_t^0 v_i(t)$ and initialization to this as

$$\psi(v_i, 0, a, c, t) = \psi(v_i - v_-, 0, a, c, t) + \sqrt{\frac{l_1}{c_1}} \psi(i_f, 0, a, c, t)$$

$$= \sqrt{\frac{l_1}{c_1}} \varphi_1(t) + \sqrt{\frac{c_2 l_1}{c_1 l_2}} \varphi_2(t).$$

Summarizing the above observation leads that zero-order gain circuit returns the input function $v_i(t)$ but provides extra time function based on initial distribution of voltage currents in the distributed elements, i.e. $\psi(v_i, 0, a, c, t)$.

8.2.8 Operational Amplifier Circuit for Semi-differintegration by Semi-infinite Lossy Line

Figure 8.8 describes the semi-infinite lossy line (a half-order element). As opposed to lossless case, the voltage current relation is of half order here as it was zero order for lossless case.

8.2 Electronics Operational Amplifier Circuits

Fig. 8.8 Semi-infinite lossy line (half-order element)

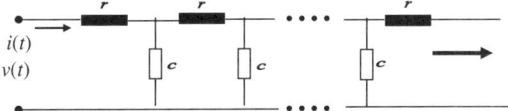

The diffusion equation corresponding to this lossy line is $\frac{\partial v(x,t)}{\partial t} = \alpha \frac{\partial^2 v(x,t)}{\partial x^2}$. In Fig. 8.8, $v(t)$ and $i(t)$ are the voltage and current at the source terminal. The per unit length distributed resistance is r and per unit length distributed capacitance is c. The diffusivity coefficient is $\alpha = rc$. The initial state of charge and voltage that exists in infinite array of elements is ψ.

The terminal characteristics is $v(t) = r\sqrt{\alpha} \frac{d^{-1/2}}{dt^{-1/2}} i(t) + \varphi_1(t)$ or

$$v(t) = r\sqrt{\alpha}_c D_t^{-1/2} i(t), \text{ where } \psi(i, -1/2, a, c, t) = \frac{1}{r\sqrt{\alpha}} \varphi_1(t) \text{ or}$$

$$i(t) = \frac{1}{r\sqrt{\alpha}} \frac{d^{1/2}}{dt^{1/2}} v(t) + \varphi_2(t) \text{ or}$$

$$i(t) = \frac{1}{r\sqrt{\alpha}}_c D_t^{1/2} v(t), \text{ where } \psi(v, 1/2, a, c, t) = r\sqrt{\alpha} \varphi_2(t).$$

8.2.9 Operational Amplifier Circuit for Semi-integrator

Figure 8.9 is the circuit for semi-integrator.

The component equations are the following:

$$v_i(t) - v_-(t) = i_i(t) R$$

$$v_-(t) - v_o(t) = r\sqrt{\alpha}_c D_t^{-1/2} i_f(t), \text{ where } \psi(i_f, -1/2, a, c, t) = \frac{1}{r\sqrt{\alpha}} \varphi_1(t).$$

Putting $i_f = i_i$ and $v_-(t) = 0$, we have the expression for semi-integration:

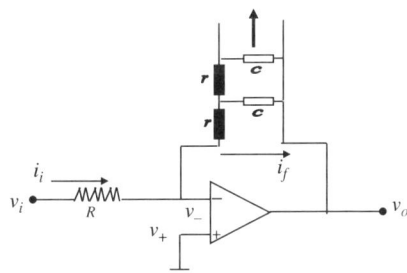

Fig. 8.9 Semi-integrator circuit (half-order integrator)

$$v_o(t) = -\frac{r\sqrt{\alpha}}{R} {_cD_t^{-1/2}} v_i(t),$$

where $\psi(v_i, t, -1/2, a, c, t) = R\psi(i_f, -1/2, a, c, t) = \frac{R}{r\sqrt{\alpha}}\varphi_1(t)$.

Alluding to Fig. 8.1, the input element and output element impedances are $Z_f = \frac{r\sqrt{\alpha}}{s^{1/2}}$ and $Z_i = R$, which leads to transfer function in s-domain as $\frac{V_0(s)}{V_i(s)} = -\frac{r\sqrt{\alpha}}{Rs^{1/2}}$. Realizing this with $r = 1K\Omega$ and $c = 1\mu F$ with $R = 22K\Omega$ gives the transfer function for the semi-integrator as $\frac{V_0(s)}{V_i(s)} = -1.4374s^{-0.5}$. Replacing $s \to j\omega$, the transfer function becomes $\frac{V_0(j\omega)}{V_i(j\omega)} = -\frac{1.4374}{\sqrt{\omega}}\exp\left(-j\frac{\pi}{4}\right)$. Following an inverting amplifier after semi-integration stage of Fig. 8.9, one gets fractional (semi) integral control block for analog fractional order PID circuit. This block will behave as a constant phase element of -45^0. The controller constant like reset rate of PID is $T_i = \frac{Z_f}{Z_i} = \frac{\sqrt{\frac{r}{c}}}{R} = 1.4374$.

8.2.10 Operational Amplifier Circuit for Semi-differentiator

Figure 8.10 shows a semi-differentiator circuit with semi-infinite lossy element

The expressions for circuit of Fig. 8.10 are $i_i(t) = \frac{1}{r\sqrt{\alpha}}{_cD_t^{1/2}}[v_i(t) - v_-(t)]$. Initializing expression is

$$\psi(v_i - v_-, 1/2, a, c, t) = r\sqrt{\alpha}\varphi_2(t)$$

and $v_-(t) - v_o(t) = i_f(t)R$. Putting $i_f = i_i$ and $v_-(t) = 0$ we have

$$v_o(t) = -i_f(t)R = -\frac{R}{r\sqrt{\alpha}} {_cD_t^{1/2}} v_i(t)$$

and $\psi(v_i, 1/2, a, c, t) = \psi(v_i - v_-, 1/2, a, c, t) = r\sqrt{\alpha}\varphi_2(t)$.

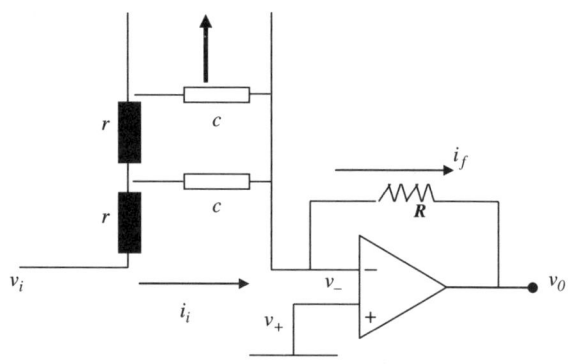

Fig. 8.10 Semi-differentiator circuit

8.2.11 Cascaded Semi-integrators

Figure 8.11 represents two semi-integrator circuits in series, which gives integral operation of order one.

The circuit of Fig. 8.11 has the following expressions:

$v_2(t) = -\frac{r_1\sqrt{\alpha_1}}{R_1} {}_cD_t^{-1/2} v_1(t)$ with $\psi_1(v_1, -1/2, a, c, t)$ is first integrator, transfer function;

$v_3(t) = -\frac{r_2\sqrt{\alpha_2}}{R_2} {}_cD_t^{-1/2} v_2(t)$ with $\psi_2(v_2, -1/2, a, c, t)$ is the second integrator expression.

Then combining we have

$$v_3(t) = \frac{r_1 r_2 \sqrt{\alpha_1 \alpha_2}}{R_1 R_2} {}_cD_t^{-1} v_1(t) \text{ with}$$

$$\psi(v_1, -1, a, c, t) = {}_cd_t^{-1/2} \psi_1(v_1, -1/2, a, c, t) - \frac{R_2}{r_2\sqrt{\alpha_2}} \psi_2(v_2, -1/2, a, c, t).$$

8.2.12 Semi-integrator Series with Semi-differentiator Circuit

Figure 8.12 gives the schematic of cascaded semi-differentiator and semi-integrator to have overall zero order.

$R_1 = R_2 = r_1 = r_2 = \sqrt{\alpha_1} = \sqrt{\alpha_2} = 1$.

The solution is $v_3(t) = -{}_cD_t^{1/2} v_2(t) = -{}_cD_t^{1/2}\left[-{}_cD_t^{-1/2} v_1(t)\right] = {}_cD_t^0 v_1(t)$, as expected. Here $v_3(t) = v_1(t)$ only if

$$\psi(v_1, 0, a, c, t) = {}_cd_t^{1/2} \psi(v_1, 1/2, a, c, t) - \psi(v_2, 1/2, a, c, t) = 0,$$

where $\psi(v_1, 0, a, c, t)$ is the combined initialization function for the combined operator ${}_cD_t^0 v_1(t)$. This is satisfied under terminal charging. However under side charging conditions, it is seen that $v_3(t) = v_1(t) + \psi(v_1, 0, a, c, t)$ and the zero property is

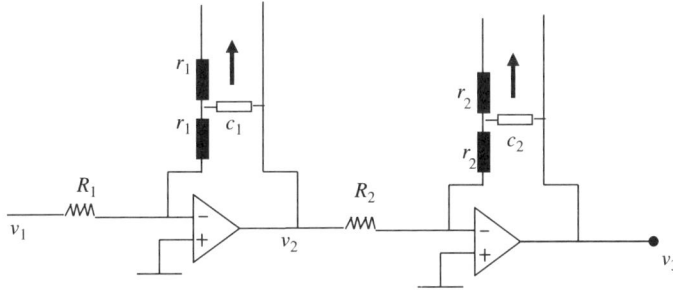

Fig. 8.11 Cascaded semi-integrators

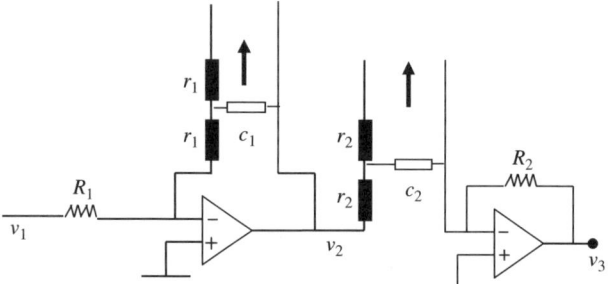

Fig. 8.12 Cascade circuit semi-integrator and semi-differentiator

not satisfied (not necessarily satisfied), as was indicated in the concept of initialized generalized calculus.

8.3 Battery Dynamics

8.3.1 Battery as Fractional Order System

This section analyzes a simple battery. Electrolytic cell is known to exhibit fractional behavior, typically of half order. The fractional system is an electrode–electrolyte interface, where diffusion takes place. This diffusion process is called the Warburg impedance (or a constant phase CPE element). The two phases of battery operation is considered: charging and discharging phases. The discharge phase is load drawing (usage). The charging phase takes place from $t = a = 0$ to $t = c$, with actual current flowing i.e. charging occurring for $a = 0 \leq t \leq b$. Later is the load drawing usage phase. Figure 8.13 gives diagram of battery circuit charging, discharging phase circuit and the charge current profile. Here constant current charging is assumed. The block "W" represents Warburg impedance of electrode–electrolyte interface.

8.3.2 Battery Charging Phase

Time domain equations from Fig. 8.13 are

$$v_1(t) - v_2(t) = \tfrac{1}{C}{}_0D_t^{-1}i_c(t) = \tfrac{1}{C}\int_0^t i_c(\tau)d\tau + v_{1-2}(0)$$
$$v_2(t) - v_3(t) = i_R(t)R$$
$$i_c(t) = i_R(t) + i_W(t), \text{ and}$$
$$i_W(t) = \tfrac{1}{B}{}_0D_t^{1/2}[v_2(t) - v_3(t)]$$

Taking Laplace transform yields

8.3 Battery Dynamics

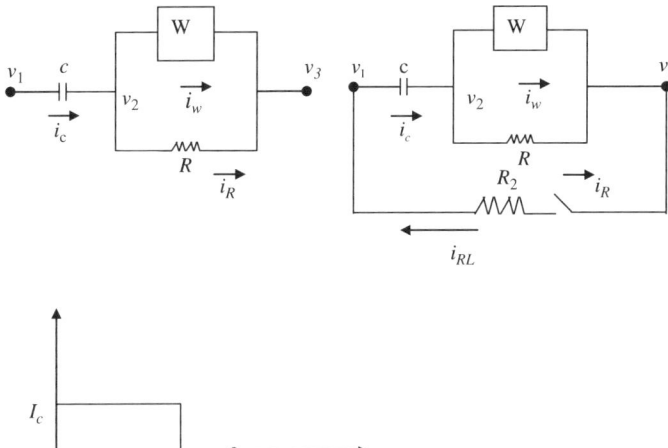

Fig. 8.13 Battery charging, discharging circuit, and charging current profile

$$V_1(s) - V_2(s) = \tfrac{1}{Cs} I_c(s) + \tfrac{v_{1-2}(0)}{s}$$
$$V_2(s) - V_3(s) = I_R(s) R$$
$$I_c(s) = I_R(s) + I_W(s) \text{ and}$$
$$I_W(s) = \tfrac{1}{B}\left[s^{1/2}(V_2(s) - V_3(s)) + \psi(v_2 - v_3, 1/2, 0, c, s)\right]$$

Eliminating $I_W(s)$, $I_R(s)$, $V_2(s)$ and rearranging, we arrive at the following:

$$I_c(s) = \frac{s\left(\tfrac{1}{B}s^{1/2} + \tfrac{1}{R}\right)}{s + \tfrac{1}{CB}s^{1/2} + \tfrac{1}{RC}} \{V_1(s) - V_3(s)\}$$
$$+ \frac{\tfrac{1}{B}s\psi(v_2 - v_3, 1/2, 0, c, s) - v_{1-2}(0)\left\{\tfrac{1}{B}s^{1/2} + \tfrac{1}{R}\right\}}{s + \tfrac{1}{CB}s^{1/2} + \tfrac{1}{RC}}$$

or
$$V_1(s) - V_3(s) = \left\{\frac{s + \tfrac{1}{CB}s^{1/2} + \tfrac{1}{RC}}{s\left(\tfrac{1}{B}s^{1/2} + \tfrac{1}{R}\right)}\right\} I_c(s) - \frac{\tfrac{1}{B}\psi(s)}{\tfrac{1}{B}s^{1/2} + \tfrac{1}{R}} + \frac{v_{1-2}(0)}{s}$$

These Laplace expressions are used to carry out transfer function analysis and to obtain time domain responses. The following are the useful Laplace transform pairs:

$$\left(\frac{B}{\sqrt{s}+\frac{B}{R}}\right) \leftrightarrow B\left[\frac{1}{\sqrt{\pi t}} - \frac{B}{R}e^{(\frac{B}{R})^2 t}erfc\left(\frac{B}{R}\sqrt{t}\right)\right]$$

$$\left[\frac{\frac{1}{C}}{\sqrt{s}\left(\sqrt{s}+\frac{B}{R}\right)}\right] \leftrightarrow \frac{1}{C}e^{(\frac{B}{R})^2 t}erfc\left(\frac{B}{R}\sqrt{t}\right)$$

$$\left[\frac{\frac{B}{RC}}{s\left(\sqrt{s}+\frac{B}{R}\right)}\right] \leftrightarrow \frac{1}{C}\left[1 - e^{(\frac{B}{R})^2 t}erfc\left(\frac{B}{R}\sqrt{t}\right)\right]$$

Using standard Laplace transform tables, the time response is got for an impulse charging ($I_c(s) = 1$) as:

$$v_1(t) - v_3(t) = \frac{1}{C} + B\left\{\frac{1}{\sqrt{\pi t}} - \frac{B}{R}\exp\left(\frac{B^2}{R^2}t\right)erfc\left(\frac{B\sqrt{t}}{R}\right)\right\},$$

During the initialization period, both the initialization functions are taken as zero for $0 = a < t \leq c$. The above time expression is for impulse charging and for any general charging, this expression is convoluted with the charging current profile i.e. general $i_c(t)$, to get following:

$$v_1(t) - v_3(t) = \int_0^t \left\{\frac{1}{C} + B\left\{\frac{1}{\sqrt{\pi \tau}} - \frac{B}{R}e^{(\frac{B}{R})^2 \tau}erfc\left(\frac{B\sqrt{\tau}}{R}\right)\right\}\right\} i_c(t-\tau)d\tau$$

As in Fig. 8.13, if the constant current charge case for charging period $i_c(t) = I_c$ for $0 = a < t \leq b$, then the step current can be resolved as function of two unit step functions. $i_c(t - \tau) = I_c(u(t - \tau) - u(t - \tau - b))$ and the charging equation thus becomes

$$v_1(t) - v_3(t) = \int_0^t \left[\frac{1}{C} + B\left\{\frac{1}{\sqrt{\pi \tau}} - \frac{B}{R}e^{(\frac{B}{R})^2 \tau}erfc\left(\frac{B\sqrt{\tau}}{R}\right)\right\}\right] I_c d\tau$$

$$- u(t-b)\int_0^{t-b} \left[\frac{1}{C} + B\left\{\frac{1}{\sqrt{\pi \tau}} - \frac{B}{R}e^{(\frac{B}{R})^2 \tau}erfc\left(\frac{B\sqrt{\tau}}{R}\right)\right\}\right] I_c d\tau$$

Using the solved integral formula

$$\int_0^t e^{E\tau}erfc\left(\sqrt{E\tau}\right)d\tau = \frac{1}{B}\left[e^{Et}erfc\left(\sqrt{Et}\right) - 1\right] + \frac{2}{\sqrt{\pi}}\sqrt{Et},$$

8.3 Battery Dynamics

we get the charging equation as:

$$v_1(t) - v_3(t) = I_c \left[\frac{t}{C} + 2B\sqrt{\frac{t}{\pi}} - \frac{R}{B} \left\{ e^{(\frac{B}{R})^2 t} erfc\left(\frac{B\sqrt{t}}{R}\right) - 1 + \frac{2}{\sqrt{\pi}} \left(\frac{B\sqrt{t}}{R}\right) \right\} \right]$$
$$- I_c u(t-b) \left[\frac{t-b}{C} + 2B\sqrt{\frac{t-b}{\pi}} - \frac{R}{B} \left\{ e^{(\frac{B}{R})^2 (t-b)} erfc\left(\frac{B\sqrt{t-b}}{R}\right) - 1 + \frac{2}{\sqrt{\pi}} \left(\frac{B\sqrt{t-b}}{R}\right) \right\} \right]$$

Figure 8.14 gives the charging voltage $v_1(t) - v_3(t)$ for charging period of $0 < t < b = 1$, and during relaxation period $t > 1$. The same figure demonstrates the voltage across Warburg element $v_2(t) - v_3(t)$. For simplicity, the parameters of battery is chosen as $R = B = C = 1.0$.

The above charging expression of total battery includes the effect of both the Warburg element and the capacitor, which is segregated as described in the following expressions. Since the $v_{1-2}(0) = 0$ must be zero, the voltage $v_2(t) - v_3(t)$ is determined as:

$$v_2(t) - v_3(t) = \{v_1(t) - v_3(t)\} - \{v_1(t) - v_2(t)\} = \{v_1(t) - v_3(t)\} - \frac{1}{C}\int_0^t i_c(\tau)d\tau$$

$$\{v_1(t) - v_3(t)\} - \frac{1}{C}\int_0^t I_c \{u(\tau) - u(\tau-b)\}d\tau$$

Using the derived expression for $v_1(t) - v_3(t)$, the Warburg element will have voltage

$$v_2(t) - v_3(t) = \int_0^t B\left\{\frac{1}{\sqrt{\pi\tau}} - \frac{B}{R}e^{(\frac{B}{R})^2 \tau} erfc\left(\frac{B\sqrt{\tau}}{R}\right)\right\} I_c d\tau$$
$$- u(t-b) \int_0^{t-b} B\left\{\frac{1}{\sqrt{\pi\tau}} - \frac{B}{R}e^{(\frac{B}{R})^2 \tau} erfc\left(\frac{B\sqrt{\tau}}{R}\right)\right\} I_c d\tau.$$

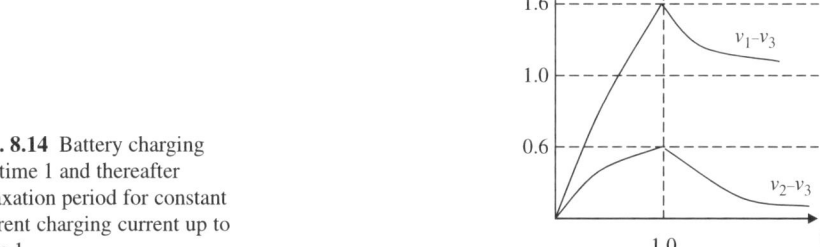

Fig. 8.14 Battery charging till time 1 and thereafter relaxation period for constant current charging current up to time 1

The above expression can be a good example for mathematical basis for an initialization function for systems of this type with constant current charging. These equations are also valid for $t > c$ with external open circuit. Also if it is desired to initialize at $c = 0$, then axis shifting with $t \to (t - c)$ may be carried out for the expression obtained for Warburg impedance.

Warburg element is a semi-derivative constitutive element

$$i_W = \frac{1}{B} {}_0D_t^{1/2} \{v_2(t) - v_3(t)\};$$

it is assumed and is terminally charged. Then the initialization function for the element is defined formally as

$$\psi(f, u, a, c, t) = \frac{d^m}{dt^m} \psi(f, -p, a, c, t) + \psi(h, m, a, c, t),$$

by standard rule described in initialization chapter. Because of assumption of terminal charging $\psi(h, 1, a, c, t) = 0$, thus $\psi_W = \frac{1}{\Gamma(1/2)} \frac{d}{dt} \int_a^c (t-\tau)^{\frac{1}{2}-1} \{v_2(\tau) - v_3(\tau)\} d\tau$, for $t > c$.

8.3.3 Battery Discharge Phase

Here in the solution, demonstration will be made to transform $s \to p$ and obtain the solution. Referring to the circuit in Fig. 8.13, we write the following:

$$v_1(t) - v_2(t) = \tfrac{1}{C} {}_cD_t^{-1} i_c(t) = \tfrac{1}{C} \int_c^t i_c(t) dt + v_{1-2}(c)$$
$$v_2(t) - v_3(t) = i_R(t)R$$
$$i_c(t) = i_R(t) + i_W(t) = i_{RL}(t)$$
$$v_3(t) - v_1(t) = i_{RL}(t)R_L \text{ and}$$
$$i_W(t) = \tfrac{1}{B} {}_cD_t^{1/2} \{v_2(t) - v_3(t)\}$$

Laplace transforming the above set, we obtain

$$V_1(s) - V_2(s) = \tfrac{1}{Cs} I_c(s) + \tfrac{v_{1-2}(c)}{s}$$
$$V_2(s) - V_3(s) = I_R(s)R$$
$$I_c(s) = I_R(s) + I_W(s) = I_{RL}(s)$$
$$V_3(s) - V_1(s) = I_{RL}(s)R_L$$
$$I_W(s) = \tfrac{1}{B} \left\{ s^{1/2} (v_2(s) - v_3(s)) + \psi_W(v_2 - v_3, 1/2, 0, c, s) \right\}$$

8.3 Battery Dynamics

From charging analysis of equation $V_1(s) - V_3(s)$ and with

$$I_c(s) = I_{RL}(s) = \{V_3(s) - V_1(s)\}/R_L$$

and taking $a = 0, c = 1$

$$V_1(s) - V_3(s) = \left[\frac{R_L s\left(\frac{s^{1/2}}{B} + \frac{1}{R}\right)}{R_L \frac{s^{3/2}}{B} + \left(1 + \frac{R_L}{R}\right)s + \frac{1}{CB}s^{1/2} + \frac{1}{RC}}\right]\left[\frac{v_{1-2}(1)}{s} - \frac{\psi_W(v_2 - v_3, 1/2, 0, 1, s)}{B\left(\frac{s^{1/2}}{B} + \frac{1}{R}\right)}\right]$$

There is no forced term (forced function is zero) but contains initialization terms from capacitor $v_{1-2}(1)$ and from Warburg impedance ψ_W, the historic past of the element.

The technique to solve the above expression is to convert $s \to p$ domain. By putting $p = s^{1/2}$, then $p^2 = s$, $p^3 = s^{3/2}$

$$V_1(p) - V_3(p) = \left[\frac{p^2\left(p + \frac{B}{R}\right)}{p^3 + \frac{B}{R_L}\left(1 + \frac{R_L}{R}\right)p^2 + \frac{1}{R_L C}p + \frac{B}{R_L RC}}\right]\left[\frac{v_{1-2}(1)}{p^2} + \frac{\frac{1}{B}\psi(v_2 - v_3, 1/2, 0, 1, p^2)}{\frac{p}{B} + \frac{1}{R}}\right]$$

Here appropriate substitution is called for $\psi(p^2)$, and then using partial fractions to achieve p domain response. This obtained p domain response may be transformed back to s domain and then inverse Laplace operation to get time domain answer. To carry out this, special transform technique is required. The special transformation is "conformal transformation."

It is possible to simplify and rewrite the above expression in p-variable as:

$$V_1(p) - V_3(p) = \frac{(p + k_1)(p^2 + k_2 p + k_3)}{(p + a)(p^3 + k_4 p + k_4)}$$

This can be partial fractioned and written as, first in p variable and then substituting $p = \sqrt{s}$ as:

$$V_1(p) - V_3(p) = \frac{A}{p+a} + \frac{B}{p+b} + \frac{C}{p+c} + \frac{D}{p+d} = \frac{A}{\sqrt{s}+a} + \frac{B}{\sqrt{s}+b} + \frac{C}{\sqrt{s}+c} + \frac{D}{\sqrt{s}+d}$$

Inverse Laplace transform will give time response as:

$$v_1(t) - v_3(t) = AF_{1/2}[-a, t] + BF_{1/2}[-b, t] + CF_{1/2}[-c, t] + DF_{1/2}[-d, t]$$

Here $F_{1/2}$ is the Robotnov–Hartley function, A, B, C, D, a, b, c, d are complex numbers (in general).

It is now possible to do fractional order system analysis and design directly on the transformed plane. To do this, it is essential to choose the greatest common fraction q of a particular system. Once this is done, all powers of s^q are replaced by powers of $p = s^q$. Then standard pole-zero analysis is done. This analysis includes root finding, partial fractions, root-locus compensation, etc. This is "conformal transformation," $s \to w$, plane is discussed in Chap. 7.

8.4 Tracking Filter

The concept of fractional calculus adds another dimension to many applications. The example of fractional order-tracking filter will demonstrate the degree of freedom in tuning. This concept is welcome in control science applications where conventional proportional integral and derivative (PID) controller gets more degree of freedom as the order control of integration and derivative part is also available. By this we get the new type of controller as $PI^\alpha D^\beta$. The fractional order-tracking filter is just one example. Here a noisy signal $x(t)$ is to be filtered to yield filtered signal $y(t)$. Filter description is

$$(_0D_t^q + a)y(t) = ax(t) \text{ or } _0d_t^q y(t) + \psi(y, q, 0, c, t) + ay(t) = ax(t)$$

Taking Laplace gives

$$s^q Y(s) + \psi(s) + aY(s) = aX(s)$$

$$Y(s) = \frac{a}{s^q + a} X(s) - \frac{\psi(s)}{s^q + a}$$

The use of $\psi(s)$ is to "pre-charge" the filter to maximize the filter lag, if taken zero then

$$\frac{Y(s)}{X(s)} = \frac{a}{s^q + a} \text{ is the transfer function of the filter.}$$

For a unit step input $X(s)$, the filter response is given by $Y(s) = \frac{a}{s(s^q+a)}$. The full range of freedom is available, $0 \le q \le 1$. Taking $q = 1/2$, we have the time response $y(t) = 1 - e^{a^2 t} erfc(at^{1/2})$.

For $q = 1/2$, the forced response is obtained for a Heaviside unit step $h(t)$ (as above), which is from the listed transform tables. Following steps demonstrate the solution of filter response for any order other than half, for unit step.

Let us write the filter output as

$$Y(s) = \frac{a}{s(s^q + a)}.$$

This we rewrite as follows:

$$Y(s) = a\frac{1/a}{s}\left(\frac{a}{s^q + a}\right) = a\frac{1/a}{s}\left(1 - \frac{s^q}{s^q + a}\right) = \frac{1}{s} - \frac{s^q}{s(s^q + a)}$$

8.4 Tracking Filter

Using the Laplace pair $L\{E_q[-at^q]\} = \frac{1}{s}[s^q L\{F_q[-at]\}]$ and

$$L\{E_q[at^q]\} = \frac{s^{q-1}}{s^q - a}, q > 0,$$

we get the inverse of Laplace of

$$Y(s) = \frac{1}{s} - \frac{s^q}{s(s^q + a)} = \frac{1}{s} - \frac{s^{q-1}}{s^q + a}$$

to get time response:

$$y(t) = [h(t) - E_q(-at^q)] = 1 - E_q[-at^q],$$

where $h(t) = \begin{cases} 1 & t \geq 0 \\ 0 & t < 0 \end{cases}$ is Heaviside step.

For a first-order RC filter (integer-order filter with order $q = 1$) the Heaviside step response is $y(t) = 1 - \exp(-at)$, with time constant $\tau = RC = 1/a$. For $q = 1$, the one-parameter Mittag-Leffler function $E_q[-at^q] = E_1[-at] = \exp(-at)$. Here, passing remark is made, as for integer order calculus, the exponential function appears as solution, for fractional order calculus Mittag-Leffler or its variants say

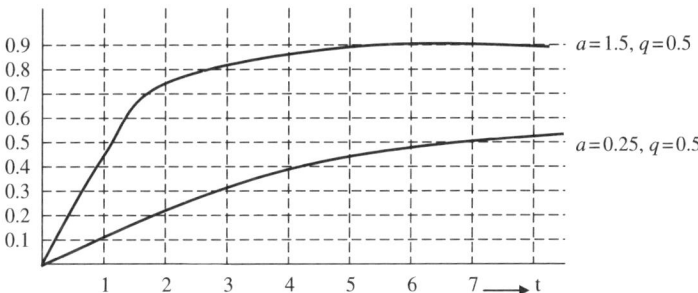

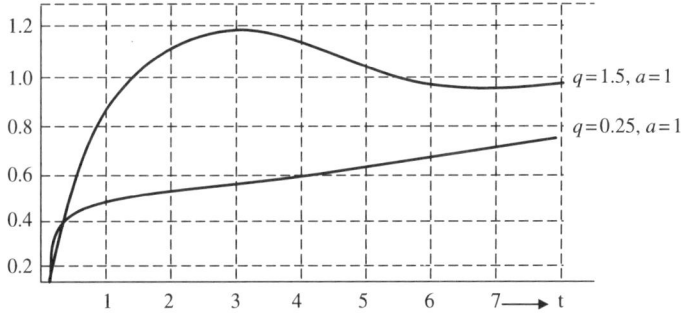

Fig. 8.15 a Response to step for varying a at $q = 0.5$; **b** Response to a step for varying q at $a = 1$

Robotnov-Hartley function, Miller-Ross function, etc. are the basis, and appears in solutions (Fig. 8.15).

8.4.1 Observations

The tracking of the filter performance with varying qth order as said in Fig. 8.16 gives a variable slope of cut-off roll over at the same corner frequency. Also varying the order q, by keeping the corner frequency unchanged, also controls the phase. Conventionally, the tuning of filter call for tuning by adjustment of a, to have particular response in time domain, whereas the extra freedom of q gives temporal response adjustment at the user's will. This extra freedom is what the fractional order tuners give i.e. to get temporal or frequency characteristics at user's will. This is the concept of "fractional pole," which is fundamental to all these fractional order systems.

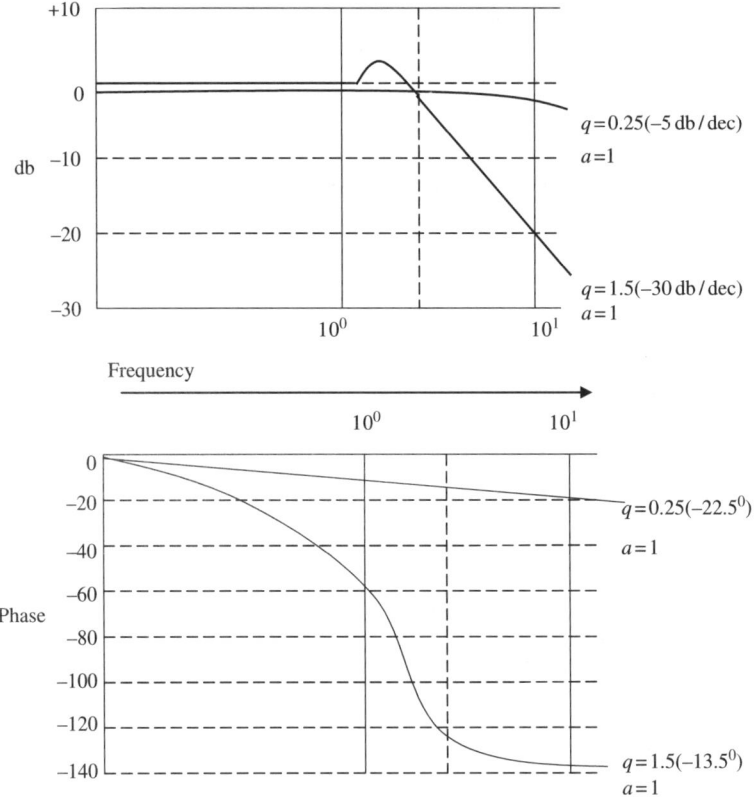

Fig. 8.16 Gain and phase plot with change of order q with constant a

8.5 Fractional Order State Vector Representation in Circuit Theory

This example, in this chapter, is chosen as a working model for vector space representation. The vector initialization issues have been touched in Chap. 7. This example will elaborate on the same topic and may be therefore extended for any system having set of fractional differential equations as basic phenomena representation. This example pertains to electrical science circuit analysis. Figure 8.17 represents the circuit diagram of a voltage source connected by an inductor to a semi-infinite lossy transmission line. As described earlier, the lossy semi-infinite line has got terminal characteristics defined by semi-differential element of fractional calculus.

Dynamic relationship between current and voltage for inductor is $v_L(t) = L_c D_t^1 i(t)$, where L is the inductance of the coil. For a semi-infinite lossy line, the dynamic relationship is $v_o(t) = \alpha_c D_t^{-1/2} i(t)$, where α is the line constant which depends on the distributed ohmic resistance per unit length and distributed capacitance per unit length. In the differential form with $c = 0$ (start time), these equations can be expressed as:

$$_0d_t^1 i(t) + \psi(1, i(t), a, 0, t) = \frac{1}{L} v_L(t) \quad\text{and}\quad _0d_t^{1/2} v_o(t) + \psi(1/2, v_o(t), a, 0, t) = \alpha i(t).$$

From Kirchoff's law, we have $v_L(t) = v_i(t) - v_o(t)$. Replacing this in the inductor expression above, we get the dynamic equation as

$$_0d_t^1 i(t) = -\frac{1}{L} v_o(t) + \frac{1}{L} v_i(t) - \psi(1, i(t), a, 0, t).$$

To have vector space of fractional dynamic variable, it is necessary to reduce all the differential relationship to differential based on the largest common (denominator) differential fraction. In this case, the choice is 1/2. We define the fractional order dynamic variable vector as:

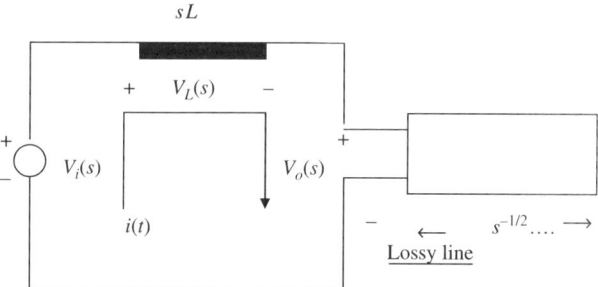

Fig. 8.17 Voltage source connected to a lossy transmission line through series inductor

$$\bar{x}(t) = \begin{bmatrix} x_1(t) \\ x_2(t) \\ x_3(t) \end{bmatrix} \equiv \begin{bmatrix} v_o(t) \\ i(t) \\ {}_0d_t^{1/2}i(t) \end{bmatrix}$$

The system input is defined as $u(t) = v_i(t)$, and the system output is chosen as $y(t) = v_o(t)$. The vector representation of the system will be as follows:

$$_0d_t^{1/2}\bar{x}(t) = \begin{bmatrix} 0 & \alpha & 0 \\ 0 & 0 & 1 \\ -\frac{1}{L} & 0 & 0 \end{bmatrix} \bar{x}(t) + \begin{bmatrix} 0 \\ 0 \\ \frac{1}{L} \end{bmatrix} \bar{u}(t) + \begin{bmatrix} -\psi(1/2, v_o(t), a, 0, t) \\ 0 \\ -\psi(1, i(t), a, 0, t) \end{bmatrix}$$

$$\bar{y}(t) = \begin{bmatrix} 1 & 0 & 0 \end{bmatrix} \bar{x}(t) + [0]\bar{u}(t)$$

Performing Laplace transformations and matrix algebraic manipulations on the above two vector space equations, we obtain

$$\bar{X}(s) = \frac{1}{s^{3/2} + \alpha/L} \begin{bmatrix} s & \alpha s^{1/2} & \alpha \\ -\frac{1}{L} & s & s^{1/2} \\ -\frac{s^{1/2}}{L} & -\frac{\alpha}{L} & s \end{bmatrix} \left\{ \begin{bmatrix} 0 \\ 0 \\ \frac{1}{L} \end{bmatrix} \bar{U}(s) - \begin{bmatrix} \psi_1(s) \\ \psi_2(s) \\ \psi_3(s) \end{bmatrix} \right\}$$

where $\psi_1(s) = \psi(1/2, v_o(t), a, 0, s)$, $\psi_2(s) = 0$ and $\psi_3(s) = \psi(1, i(t), a, 0, s)$, which is constant for an inductor. The Laplace transform of the forced response $\bar{X} F(s)$ is the first term, from the above relation is

$$\bar{X} F(s) = \frac{1/L}{s^{3/2}+\alpha/L} \begin{bmatrix} \alpha \\ s^{1/2} \\ s \end{bmatrix} \bar{U}(s). \text{ The Laplace transform of the initialized}$$

response $\bar{X}_i(s)$ is from the second term that is:

$$\bar{X}_i(s) = \frac{-1}{s^{3/2} + \alpha/L} \begin{bmatrix} s\psi_1(s) + \alpha s^{1/2}\psi_2(s) + \alpha\psi_3(s) \\ \left(\frac{-1}{L}\right)\psi_1(s) + s\psi_2(s) + s^{1/2}\psi_3(s) \\ \left(\frac{-s^{1/2}}{L}\right)\psi_1(s) + \left(\frac{-\alpha}{L}\right)\psi_2(s) + s\psi_3(s) \end{bmatrix}$$

These expressions can now be evaluated for any specific input and initialization function.

The choice of fractional dynamic variable order basis q should imply minimal configurations. In the above example choosing $q = 1/4$ rather than $1/2$ would yield six dynamic variables, instead of three. It is also important to remember that the least number of fractional dynamic variables is obtained by choosing the basis q as the largest common (denominator) fraction of the differential order (for $q_1 = 1/2$ and $q_2 = 1/3$ would require basis $q = 1/6$).

The input output transfer function with $L = 1$ is

$$G(s) = \frac{V_0(s)}{V_i(s)} = \frac{\frac{1}{\sqrt{s}}}{s + \frac{1}{\sqrt{s}}} = \frac{1}{s^{3/2} + 1}$$

8.5 Fractional Order State Vector Representation in Circuit Theory

The time domain representation of the transfer function (without initialization) is thus the following:

$$_0d_t^{3/2} v_0(t) + v_0(t) = v_i(t).$$

Here initialization is zero,

$$\psi(v_0, 3/2, a, 0, t) = 0,$$

thus, operation is $_0D_t^{3/2} v_0(t) = {_0d_t^{3/2}} v_0(t)$. This problem can be solved in several ways depending upon the specific input. Using $F_q[a, t]$ function as obtained for the solution of "fundamental fractional order differential equation," the impulse response of this circuit with $q = 3/2$ as basis, and $v_i(t) = \delta(t)$, $V_i(s) = 1$, the solution is

$$v_0(t) = L^{-1}\left\{\frac{1}{s^{3/2}+1}\right\} = F_{3/2}[-1.t].$$

For a Heaviside step $h(t) = \begin{cases} 1 & t \geq 0 \\ 0 & t < 0 \end{cases}$ input $V_i(s) = 1/s$, the solution is

$$v_0(t) = L^{-1}\left\{\frac{1}{s\left(s^{3/2}+1\right)}\right\} = h(t) - E_{3/2}\left[-t^{3/2}\right]$$

For the same circuit, solution basis is now chosen to be $q = 1/2$. Here w-plane "conformal transformation" is demonstrated. This transformation was described in Chap. 7, and let $s^{1/2} = w$. Putting this, we get w-plane transfer function as:

$$\frac{V_0(w)}{V_i(w)} = G(w) = \frac{1}{w^3+1}$$

The denominator $w^3 + 1$ has one root at $w = -1$. Dividing the polynomial $w^3 + 1$ by $(w + 1)$ gives the other factor as, $w^2 - w + 1$. The other two roots of this are from roots of $w^2 - w + 1$, those are $w = \frac{1}{2} \pm j\frac{\sqrt{3}}{2} = e^{\pm j\pi/3}$. Therefore, the factorized form of denominator is

$$w^3 + 1 = (w - \{-1\})\left(w - \left\{\frac{1}{2} + j\frac{\sqrt{3}}{2}\right\}\right)\left(w - \left\{\frac{1}{2} - j\frac{\sqrt{3}}{2}\right\}\right)$$

This transfer function has poles in w-plane at $w = -1$, $w = e^{+j\pi/3}$, and $w = e^{-j\pi/3}$. Since the basis $q = 1/2$, the stability wedge is at angle

$$\phi = \pm q\frac{\pi}{2} = \pm\frac{\pi}{4}$$

. All these poles are to the left of the instability wedge in w-plane. The two poles $\exp(+j\pi/3)$ and $\exp(-j\pi/3)$ in the right half of the w-plane correspond to poles at $s = e^{+j2\pi/3}$ and $s = e^{-j2\pi/3}$ in s-plane, therefore an oscillatory response is suggested. The third pole in w-plane $w = -1$ is in "hyper- damped" zone. This pole adds a rapidly decaying time response, to the oscillatory response due to the other two poles.

Expanding the $G(w)$ obtained for base $q = 1/2$, we get the following:

$$G(w) = \frac{1}{w^3+1} = \frac{1}{3}\left\{\frac{1}{w+1} - \frac{\frac{1}{2}+j\frac{\sqrt{3}}{2}}{w-\left(\frac{1}{2}+j\frac{\sqrt{3}}{2}\right)} - \frac{\frac{1}{2}-j\frac{\sqrt{3}}{2}}{w-\left(\frac{1}{2}-j\frac{\sqrt{3}}{2}\right)}\right\}$$

$$= \frac{0.3333}{w+1} - \frac{0.1667+j0.2887}{w-(0.5+j0.866)} - \frac{0.1667-j0.2887}{w-(0.5-j0.866)}$$

$$= \frac{0.333}{s^{1/2}+1} - \frac{0.1667+j0.2887}{s^{1/2}-(0.5+j0.866)} - \frac{0.1667-j0.2887}{s^{1/2}-(0.5-j0.866)}$$

The time response can be obtained by inverse Laplace transforming and by using

$L\{F_q[\pm a, t]\} = \frac{1}{s^q \mp a} = \frac{1}{w \mp a}$ pair as:

$$v_0(t) = 0.3333 F_{1/2}[-1, t] - (0.1667 + j0.2887) F_{1/2}[0.5 + j0.866, t]$$
$$- (0.1667 - j0.2887) F_{1/2}[0.5 - j0.866, t]$$

$v_0(t)$ thus obtained above for the circuit of semi-derivative terminal point immitance, connected to a voltage source through an inductor, is for only "forced-response." Similarly, the solution of the "initialization response" may be obtained and then superimposed on this.

8.6 Concluding Comments

Here, a variety of electrical and electronic circuits have given the feel of generalized fractional calculus, approaches. The intentions are to demonstrate the broad array of uses of the fractional calculus, to clearly delineate the effects of the initialization function, to contrast generalized versus integer order differentiation and integration, to demonstrate the generalized "zero operation" (inverse operation), to demonstrate the use of Laplace transform methods, to show some unusual aspects of mathematics concerning the modeling of distributed effects—broadly to give scientists and engineers a concept to model the reality of system being investigated by use of fractional calculus.

Chapter 9
Application of Generalized Fractional Calculus in Other Science and Engineering Fields

9.1 Introduction

In this chapter, a series of scientific and engineering application is shown, where the fractional calculus is finding application. We start with diffusion model in electrochemistry, electrode–electrolyte interface, capacitor theory, fractance circuits, application in feedback control systems, viscoelasticity, and vibration damping system. This survey cannot cover the complete applications like modern trends in electromagnetic theory, such as fractional multipole, hereditary prediction of gene behavior, fractional neural modeling in bio-sciences, communication channel traffic models, chaos theory; hence simple applications are provided for appreciation. However, in Sect. 9.6, attempt is made to provide vector state feedback controller and observer that is available for multivariate control science.

9.2 Diffusion Model in Electrochemistry

One of the important studies in electrochemistry is the determination of concentration of analyzed electroactive species near the electrode surface. The characteristic describing function is found experimentally as $m(t) = {}_0D_t^{-0.5}i(t)$, which is the fractional (half) integral of the current. Then the subject of interest is to find surface concentration $C_s(t)$ of the electro active species, which can be evaluated as $C_s(t) = C_0 - k({}_0D_t^{-0.5}i(t))$, where $k = 1/(nAF\sqrt{D})$—A being electrode area, n number of electrons involved in the reaction, D is the diffusion coefficient, and F is the Faraday constant. C_0 is the uniform concentration of the electro active species throughout the electrolyte medium, at the initial equilibrium situation characterized by constant potential at which the electrochemical reaction is possible.

The relationship is derived from the classical diffusion equation:

$$\frac{\partial C(x,t)}{\partial t} = D\frac{\partial^2 C(x,t)}{\partial x^2} \text{ for } (0 < x < \infty \& t > 0)$$

$$\text{with } C(\infty, t) = C_0 \& C(x, 0) = C_0 \text{ and } \left[D\frac{\partial C(x,t)}{\partial x}\right]_{x=0} = \frac{i(t)}{nAF}$$

S. Das, *Functional Fractional Calculus for System Identification and Controls.*
© Springer 2008

(similar equation for lossy semi-infinite transmission line and heat flux studies)
Some interesting points are listed below:

(1) $m(t)$ is characteristic intermediate between the current $i(t)$ and the charge passed $q(t)$. The charged passed is an integral $q(t) = {}_0D_t^{-1}i(t)$. This hints at non-conservation law of charges, as $m(t)$ manifests.
(2) The kinetics of the electrode process and the surface property of the electrode (alluding to heterogeneity) are not assumed.
(3) Instead of classical diffusion equation, it is possible to model with fractional order diffusion equation as:

$$_0D_t^\alpha C(x,t) = D\frac{\partial^2 C(x,t)}{\partial x^2}$$

with $0 < \alpha < 1$, then the surface concentration will be related to $m_\alpha(t) = {}_0D_t^{\alpha/2}i(t)$.

9.3 Electrode–Electrolyte Interface Impedance

Warburg impedance in electrical battery (cell) is another motivating example of the reality of fractional calculus. Limitation of electrical batteries, which always exhibit a limited current output, is due to the fact that microscopic electrochemical process at the electrode–electrolyte interface has "finite rate" and limits the current output. Battery manufacturers use porous electrodes to circumvent this limitation, by way of increasing the surface area. It has been experimentally established that metal–electrolyte surface interface impedance does not exhibit pure capacitance behavior, instead governed as power law: $Z(\omega) = K/(j\omega)^\alpha$, where $0 < \alpha < 1$. In Laplace term, $Z(s) = Ks^{-\alpha}$. This power term approaches unity (impedance tending to capacitive) as the smoothness of the interface is increased to infinity.

Warburg impedance $Z(\omega) \propto (j\omega)^{-0.5}$, for any solid-state diffusion, electrochemistry, gives rise to power law in frequency. The constitutive equation is

$$\frac{\partial C}{\partial t} = D\frac{\partial^2 C}{\partial x^2},$$

which gives rise to $Z(\omega) \propto (j\omega)^{-0.5}$. When diffusion takes place in a layer of thickness L driven by diffusion over voltage at $x = 0$, the observed behavior is not solely Warburg impedance. In a spatially restricted situation, there are at least two domains in the impedance spectra, which are separated by characteristic frequency $\omega_d = D/L^2$. Warburg impedance occurs at a high-frequency regime, $\omega > \omega_d$. At low frequency regime $\omega < \omega_d$, the impedance behavior depends on whether the diffusion species are reflected or extracted at the end of region ($x = L$). The reason for this is that the frequency ω_d corresponds to the transit time for a diffusing particle injected at $x = 0$, to cover a distance L. For $\omega > \omega_d$, the particles will not sense the

9.3 Electrode–Electrolyte Interface Impedance

boundary $x = L$, so the system will behave as "semi-infinite" media. Anomalous diffusion is characterized by a mean square displacement of the diffusing particles that does not follow the ordinary linear law $< r^2 > \propto t$, but more generally $< r^2 > \propto t^\beta$.

Not surprisingly, many different mechanisms give rise to anomalous behavior—like complex flow, flow through porous, flow through random tubular ion exchange resin—due to structural complexity, shapes of complex geometry, etc.

The continuity equation is the fundamental conservation law

$$\frac{\partial C}{\partial t} = -\frac{\partial J}{\partial x}$$

relating to time variation of the number density C of the diffusing species to the macroscopic flux, J. The constitutive equation for diffusion is $J = -D(\partial C/\partial x)$. To calculate diffusion impedance, consider $E(at\ x = 0) = (dE/dC)_{at x=0} C(at\ x = 0)$ where E is the excitation voltage $at\ x = 0$; the expression gives linear dependence at $x = 0$. The current is $i(at\ x = 0) = qAJ$. Taking the Laplace as the above expressions of voltage and currents are for small changing quantities with respect to time, we obtain at $x = 0$, $E(s) = (dE/dC)_0 C(s)$ and $I(s) = qAJ(s)$. Thus the diffusion impedance is $Z(s) = \frac{E(s)}{I(s)} = R_W \frac{D}{L} \frac{C(s)}{J(s)}$ where Warburg constant is R_W and is given by $R_W = \frac{L}{qAD} \left(\frac{dE}{dC}\right)_0$.

The various diffusion equations for small amplitude sinusoidal concentration may be written as in the Laplace domain form by taking Laplace of continuity and constitutive equations: $sC(s) = -\frac{\partial J(s)}{\partial x}$ and $J(s) = -D\frac{\partial C(s)}{\partial x}$. From these, we get $\frac{\partial^2 C}{\partial x^2} = \frac{1}{\lambda^2} C(s)$.

Here

$$\lambda(s) = \sqrt{\frac{D}{s}} = \sqrt{\frac{\omega_d}{s}} L$$

is the function of frequency. This Laplace (time–frequency) equation has the most general solution as $C(s) = B_1 \cosh \frac{x}{\lambda} + B_2 \sinh \frac{x}{\lambda}$.

The diffusion impedance at origin for a semi-infinite case having boundary condition $C(x = L) = 0$ is

$$Z(s) = R_W \frac{D}{L} \frac{C(s)}{J(s)} = R_W \left(\frac{\omega_d}{s}\right)^{1/2} \tanh\left[\left(\frac{s}{\omega_d}\right)^{1/2}\right].$$

The same can be represented as semi-infinite lossy transmission line model too with r and c, as per the unit length resistance and capacitance expressed as $r = R_W/L$ and $c = L/R_W D$.

9.4 Capacitor Theory

Capacitor is a charge storage devise and it is assumed that whatever charges are pumped they are held between the plates (electrodes) by ideal dielectric, having no loss. Therefore, the impedance of non-leaky capacitor or ideal capacitor is $Z_c = (1/j\omega C)$, with the dielectric constant assumed to be ideal as $\varepsilon = \varepsilon' - j0$. Now if we really have dielectric absorption, then the real capacitor is $Z_c = 1/(j\omega C)^{0.999}$, having slight imperfection, but good for electronic circuits and sample and hold circuits, etc. Now if one designs a leaky capacitor by selecting dielectric (not ideal) but having $\varepsilon = \varepsilon' - j\varepsilon''$ with $\varepsilon' = \varepsilon'' = 10^6$ for wide range of frequency and temperature then $\varepsilon = \varepsilon_r \omega^{-1/2}(1-j) = \varepsilon_r \sqrt{2}(j\omega)^{-1/2}$. Using this dielectric, the capacitance impedance is

$$Z = \frac{1}{j\omega C \varepsilon} = \frac{1}{j\omega C \varepsilon_r \sqrt{2}(j\omega)^{-1/2}} = \frac{1}{C \varepsilon_r \sqrt{2}(j\omega)^{1/2}}.$$

The dielectric is lithium hydrazium sulphate (LiN$_2$H$_2$SO$_4$). These capacitors are useful to realize the analog circuits for fractional order controller. Figure 9.1 gives the operational amplifier realization of using these capacitors.

This fact was experimented in the late nineteenth century by M.J. Curie, who noted the current voltage relation as (power law) $i(t) = V/ht^v$, where ($0 < v < 1, t > 0$). In this expression, h is related to the capacitance of the capacitor and the kind of dielectric, and v is related to losses of the capacitor. The transfer function of this model of capacitor is found to be $H(s) = C_\phi s^v$, where C_ϕ is model constant close to what is called the capacitance. The capacitor impedance is described by $Z(s) = 1/C_\phi s^v$. Advance research has realized that these half-order capacitors by polymer-coated electrode have been used as a constant phase element to get semi-differentigration circuits.

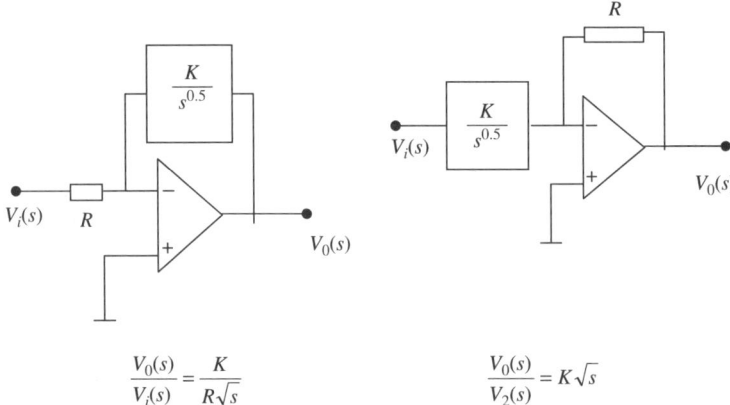

Fig. 9.1 Operational amplifier circuit to have fractional integration and differentiation

9.5 Fractance Circuit

Electrical circuit related to fractional calculus is fractance, an electrical circuit behaving in between capacitance and resistance. An example of fractance is tree fractance shown in Fig. 9.2, a self-similar structure.

The impedance of tree fractance is

$$Z(j\omega) = \sqrt{\frac{R}{C}} \frac{1}{\sqrt{\omega}} \exp\left(\frac{-j\pi}{4}\right);$$

this corresponds to the fractional order transfer function

$$Z(s) = \sqrt{\frac{R}{C}} \frac{1}{\sqrt{s}}.$$

A circuit exhibiting fractional order behavior is called fractance, not essentially limited to half order, as described in the self-similar circuit diagram of Fig. 9.2. The order can be of any arbitrary order in general. The fractance devises have the following properties. They are constant phase elements, i.e. the phase angle is constant, independent of the the frequency within wide range of frequency band. Second it is possible to construct a filter, which has moderate characteristics, which cannot be realized by using conventional devises. Generally speaking there are three basic fractance devises. The most popular one is domino ladder circuit network. Very often these are used as binary tree structure, as in Fig. 9.2. Also balanced transmission line structure is used (or symmetrical domino ladder). Design of fractances can be easily done by any of these topological configuration as mentioned, to realize the rational approximated transfer function for the fractional order Laplace operator (Chap. 7). Truncated continued fraction expansion (CFE) does not require any further transformation, a rational approximation based on any other method as (say Newton method of Carlson, described in Chap. 7) must be transformed into the form of CFE. The values of the electrical circuit elements, which are necessary for realizing a fractance, are then determined from the obtained continued fraction. If

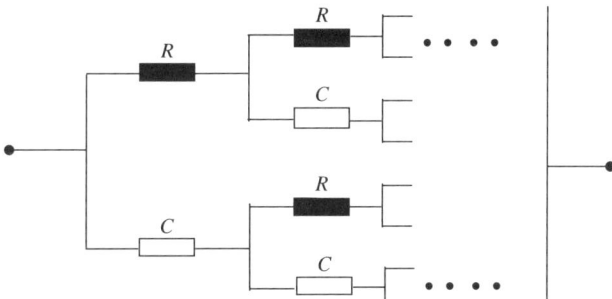

Fig. 9.2 Tree fractance circuit

all the coefficients of obtained CFE are positive, then the fractance can be simply made by passive elements (R, L, C). If some of the coefficients are negative, then the fractance realization requires active circuit as negative impedance converters, realized by operation amplifier circuits. In some of the methods of CFE to realize the represented transfer function, negative impedances do appear. Negative impedance converters are also called current inverter and have transfer function as $V_i/I_i = -Z$, realized by operational amplifier.

Recalling the semi-infinite lossy transmission line, where r is series resistance per unit length, and c is the shunt capacitance per unit length, used to demonstrate semi-differentigration as derived terminal impedance (Chap. 3). The same circuit, truncated, if we mentally reconstruct the equivalent impedance by traveling from right side of the transmission line towards the left side, we will get the following:

$$Z_{rc}(s) = r + \cfrac{1}{cs + \cfrac{1}{r + \cfrac{1}{cs + \cfrac{1}{\ddots \cfrac{1}{cs + \cfrac{1}{r + \frac{1}{cs}}}}}}}$$

For the shunt element, the admittance is $Y(s) = cs$, and thus impedance is $Z(s) = 1/Y(s)$. This truncated $Z_{rc}(s)$ is the approximation to the driving point impedance, which is $\sqrt{r/cs}$ (fractional Laplace operator) for semi-infinite line. The generalization of this with series impedances of different values, $Z_1(s), Z_3(s)\ldots Z_{2n-3}(s), Z_{2n-1}(s)$, and shunt admittances of different values, $Y_2(s), Y_4(s)\ldots Y_{2n-2}(s), Y_{2n}(s)$, gives the CFE form of domino ladder immitance as:

$$Z(s) = Z_1(s) + \cfrac{1}{Y_2(s) + \cfrac{1}{Z_3(s) + \cfrac{1}{\ddots \cfrac{1}{Y_{2n-2}(s) + \cfrac{1}{Z_{2n-1}(s) + \frac{1}{Y_{2n}(s)}}}}}}$$

The transfer function $Z(s) = \frac{s^2 + 4s + 3}{s^3 + 6s^2 + 8s}$ is realized by doing CFE for admittance as:

$$Y(s) = \frac{s^3 + 6s^2 + 8s}{s^2 + 4s + 3} = s + \cfrac{1}{\frac{1}{2} + \cfrac{1}{\frac{4}{3}s + \cfrac{1}{\frac{3}{2} + \frac{1}{\frac{1}{3}s}}}},$$

Meaning, shunt capacitors are $1, \frac{4}{3}, \frac{1}{3}$, corresponding to admittances, $Y_2(s), Y_4(s), Y_6(s)$, the series impedance in this case is $Z_1(s) = 0$, and the other series impedances corresponding to $Z_3(s), Z_5(s)$ are $\frac{1}{2}, \frac{3}{2}$. This is standard Caurer-I form of circuit synthesis. The same can be realized as Caurer-II form. Depending on the rational polynomial, the realization with CFE can have R, C, L components, combination of them, or even negative impedances, in the ladder form.

The fundamentals of circuit synthesis are applicable for this fractance realization, of rational transfer function synthesis for fractional order Laplace operator.

9.6 Feedback Control System

Feedback control system is one of the major areas where the concept of fractional calculus should be applied to obtain efficient system. This concept gives an overall efficiency (in terms of energy), also longevity and freedom, to the control engineer to compensate any shifts in the transfer function due to parametric spreads aging, etc. A system is efficient if the controller were of similar order to that of a plant (system) being controlled. In reality, the systems are of fractional order and therefore to have a fractional order controller will be efficient. Even for integer order systems, the fractional controls give better freedom to achieve what is "iso-damping."—Meaning, to achieve an overall close loop behavior of overshoot that is independent of feed forward gain (payload, amplifier feed forward gain, in power systems the load current/load resistance). H.W. Bode envisaged this concept of having fractional integrator circuits to achieve overshoot independent of the amplifier gain in 1945. He proposed a fractional order controller, the purpose of which is to have a feedback amplifier of good linearity and stable gains, even though the amplifier show non-linear characteristics and variable gain over ambience and time. Bode proposed a feedback amplifier, whose open loop frequency characteristics $G_0(j\omega)$ is such that the gain is constant for $0 < \omega < \omega_0$ and phase is constant or $-\pi(1-y)$ radians for $\omega > \omega_0$. The suggested value was $y = 1/6$, which guarantees a phase margin (PM) of $30°$. The open loop transfer function is given as

$$G_0(j\omega) = \frac{A_0}{\left[\sqrt{1-(\omega/\omega_0)^2} + j(\omega/\omega_o)\right]^{2(1-y)}},$$

meaning $|G_0(j\omega)| = A_0$ for $\omega < \omega_0$ and angle i.e. $\arg G_0(j\omega) = -\pi(1-y)$ radians for $\omega > \omega_0$. This is early development of fractional order controls. Thus it was recognized that the open-loop transfer function of a good control system shows a fractional order integral form with a fractional order between 1 and 2 (between totally being first order and second order), meaning that open-loop transfer function should be like $G_0(s) = 1/s^k$. This gives close loop transfer function as $G_{CL}(s) = \frac{1}{1+G_0(s)} = \frac{1}{s^k+1}, s = j\omega.$

In close loop transfer function $G_{CL}(s) = \frac{1}{s^k+1}$ expression, put for $s = j\omega$, $j = \cos\frac{\pi}{2} + j\sin\frac{\pi}{2}$ then

$$s = \omega\left[\cos\frac{\pi}{2} + j\sin\frac{\pi}{2}\right]$$

$$s^k = \omega^k\left[\cos\frac{k\pi}{2} + j\sin\frac{k\pi}{2}\right] = \omega^k\cos\frac{k\pi}{2} + j\omega^k\sin\frac{k\pi}{2},$$

Put this value of ω^k in $G_{CL}(s)$ to get

$$G_{CL}(s) = \frac{1}{s^k+1} = \frac{1}{\omega^k\cos\frac{k\pi}{2} + j\omega^k\sin\frac{k\pi}{2} + 1} = \frac{1}{(\omega^k\cos\frac{k\pi}{2} + 1) + j\omega^k\sin\frac{k\pi}{2}}$$

$$|G_{CL}(s)| = \frac{1}{\left[\omega^{2k}\cos^2\frac{k\pi}{2} + 1 + 2\omega^k\cos\frac{k\pi}{2} + \omega^{2k}\sin^2\frac{k\pi}{2}\right]^{0.5}}$$

$$= \frac{1}{\sqrt{\left[\omega^{2k} + 2\omega^k\cos\frac{k\pi}{2} + 1\right]}}$$

M_r is the maximum value of $|G_{CL}(s)|$ at ω_r when the denominator $\omega^{2k} + 2\omega^k\cos\frac{k\pi}{2} + 1$ is minimum.

Therefore $\frac{d}{d\omega}\left[\omega^{2k} + 2\omega^k\cos\frac{k\pi}{2} + 1\right] = 0$ gives $2k\omega^{2k-1} + 2k\omega^{k-1}\cos\frac{k\pi}{2} = 0$, meaning at $\omega^k = -\cos\frac{k\pi}{2}$ the magnitude of is maximized.

$\omega_r = \left|\cos\frac{k\pi}{2}\right|^{1/k}$ and putting this value of $\omega = \omega_r$ in expression of $|G_{CL}(s)|$, we get

$$M_r = \frac{1}{\sqrt{\left[\left(-\cos\frac{k\pi}{2}\right)^2 + 2\left(-\cos\frac{k\pi}{2}\right)\cos\frac{k\pi}{2} + 1\right]}} = \frac{1}{\sqrt{\cos^2\frac{k\pi}{2} - 2\cos^2\frac{k\pi}{2} + 1}}$$

$$= \frac{1}{\sqrt{1 - \cos^2\frac{k\pi}{2}}} = \frac{1}{\sin\frac{k\pi}{2}}$$

For finding the damping ratio, we find the poles of $G_{CL}(s)$ by transformation to w-plane and then with respect to s plane we look at the pole location.

Putting $w = s^k$ in the expression of close loop transfer function, we obtain $G_{CL}(w) = \frac{1}{w+1}$ with poles at $w = 1e^{\pm j\pi}$ in w-plane.

Therefore, the s-plane pole is at $w^{1/k}$, meaning poles at $s = (1)^{1/k}e^{\pm j\pi/k}$ in the s-plane. The line with angle $(\pm\pi/k)$ with the positive real axis of the s-plane is the locus of poles for $G_{CL}(s)$ and are called iso-damped lines for particular value of k. The damping ratio $\varsigma = \frac{\Re e(s)}{|s|}$ with respect to imaginary $(j\omega)$ axis.

The angle of the iso-damped line with respect to imaginary axis is $\left(\frac{\pi}{k} - \frac{\pi}{2}\right)$ and thus anywhere on this line the pole be, the damping ratio is

$$\varsigma = \frac{\Re e(s)}{|s|} = \sin\left(\frac{\pi}{k} - \frac{\pi}{2}\right).$$

9.6 Feedback Control System

This close-loop transfer function gives step response properties of controlled system output as robustness and stability measures.

$$|G_{CL}(s)| = \left| \frac{1}{[\omega^k \cos(k\pi/2) + 1 + j\omega^k \sin(k\pi/2)]} \right|$$

$$= \frac{1}{[\omega^{2k} + 2\omega^k \cos(k\pi/2) + 1]^{0.5}}$$

$$M_r = \frac{1}{\sin(k\pi/2)}, \quad \omega_r = |\cos(k\pi/2)|^{1/k}.$$

The amplitude takes the peak value M_r at ω_r. The damping ratio can be obtained from the poles of $G_{CL}(s)$ as $\varsigma = \sin\left(\frac{\pi}{k} - \frac{\pi}{2}\right)$. The phase margin is given by $PM = \pi - k\frac{\pi}{2} = 90(2-k)^0$. The overshoot can be expressed as the approximate formula as $M_P \cong (k-1)(0.8k - 0.6)$ per unit. These performance specifications are also termed as robustness measures and are listed for various fractional orders (Table 9.1).

General properties of Bode's ideal control system transfer functions are

a. Open loop:

   Type: $G_0(s) = \frac{K}{s^k}, (1 < k < 2)$
   Magnitude: constant slope of $-k20\,dB/dec$
   Cross-over frequency: a function of gain K
   Phase: horizontal line of $-k\pi/2$
   Nyquist: straight line at argument $-k\pi/2$

b. Closed loop:

   Type: $G_{CL}(s) = \frac{K}{s^k + K}, (1 < k < 2)$
   Gain margin: $A_m = \infty$ infinite
   Phase margin: $\Phi_m = \pi\left(1 - \frac{k}{2}\right)$, constant
   Step response: $y(t) = Kt^k E_{k,k+1}(-Kt^k)$

Table 9.1 Robustness measures for various fractional orders k

k (order)	PM (degree)	ς	$M_P\%$	M_r
1	90	1	0	1
1.1	81	0.96	2.8	1.0125
1.2	72	0.87	7.4	1.0515
1.3	63	0.75	13.6	1.1223
1.4	54	0.62	21.1	1.2361
1.5	45	0.5	30.0	1.4142
1.6	36	0.38	40.5	1.7013
1.7	27	0.27	52.5	2.2027
1.8	18	0.17	66.3	3.2361
1.9	9	0.083	82.1	6.3925

The solution to this is as follows.

The step response of a system with closed loop transfer function $G_{CL}(s)$ is

$$Y(s) = \frac{1}{s} G_{CL}(s).$$

The closed loop transfer function is $G_{CL}(s) = \frac{K}{s^k + K}$.

Therefore:

$$Y(s) = \frac{1}{s} G_{CL}(s) = \frac{K}{s(s^k + K)}$$

$$Y(s) = K \frac{1/K}{s} \left(\frac{K}{s^k + K}\right) = K \frac{1/K}{s} \left[1 - \frac{s^k}{s^k + K}\right] = \frac{1}{s} - \frac{s^k}{s^k + K}$$

$$= \frac{1}{s} - \frac{s^{k-1}}{s^k + K}$$

Inverse Laplace transforming the above, we get

$$y(t) = \left[1 - E_k(-Kt^k)\right]$$

Using series expansion of one-parameter Mittag-Leffler function, we obtain

$$y(t) = 1 - \left[1 + \frac{(-Kt^k)}{\Gamma(k+1)} + \frac{(-Kt^k)^2}{\Gamma(2k+1)} + \frac{(-Kt^k)^3}{\Gamma(3k+1)} + \ldots\right]$$

$$= -\frac{(-Kt^k)}{\Gamma(k+1)} - \frac{(-Kt^k)^2}{\Gamma(2k+1)} - \frac{(-Kt^k)^3}{\Gamma(3k+1)} - \ldots$$

$$= \frac{(Kt^k)}{\Gamma(k+1)} - \frac{(Kt^k)^2}{\Gamma(2k+1)} + \frac{(Kt^k)^3}{\Gamma(3k+1)} - \ldots$$

$$= Kt^k \left[\frac{1}{\Gamma(k+1)} + \frac{(-Kt^k)}{\Gamma(2k+1)} + \frac{(-Kt^k)^2}{\Gamma(3k+1)} + \ldots\right]$$

Now using the definition of two-parameter Mittag-Leffler function

$$E_{\alpha,\beta}(z) = \sum_{n=0}^{\infty} \frac{z^n}{\Gamma(\alpha n + \beta)},$$

we can write

$$E_{k,k+1}(-Kt^k) = \frac{1}{\Gamma(k+1)} + \frac{(-Kt^k)}{\Gamma(2k+1)} + \frac{(-Kt^k)^2}{\Gamma(3k+1)} + \ldots$$

and use this to have time response expression as:

$$y(t) = Kt^k E_{k,k+1}(-Kt^k)$$

9.6 Feedback Control System

The solution with one-parameter Mittag-Leffler and two-parameter Mittag-Leffler are equivalent. More form of algebraic manipulations of the series can give different compact functions as solutions (say by Robotnov–Hartley and several other variants of Mittag-Leffler).

Therefore for Heaviside step input $h(t) = \begin{cases} 1 & t \geq 0 \\ 0 & t < 0 \end{cases}$, the time response of output is

$$y(t) = Kt^k E_{k,k+1}(-Kt^k) = 1 - E_k(-Kt^k) = h(t) - E_k(-Kt^k)$$

Furthermore, if the composition of a feedback controller with fractional differentiator and α-order fractional integrator gives $k = 1 + \alpha$, while obtaining $G_{CL}(s)$, then:

$$y(t) = Kt^{1+\alpha} E_{1+\alpha, 2+\alpha}(-Kt^{1+\alpha})$$

Now with especially digital and advanced analog techniques, this fractional order control system is possible in having real time responses. Figure 9.3 shows the concept of fractional order PID system connection.

In fractional order PID control what is extra freedom to operator is in terms of the two extra knobs, namely in the order of differentiation and the order of integration. In these fractional order controller $PI^\lambda D^\delta$, the knob values λ, δ are between 0 and 1. In PID type control, we compensate only the dominant roots and with these extra freedoms we are able to continuously span the area shaded in Fig. 9.4. The Laplace domain of this controller has the form $H(s) = \frac{U(s)}{E(s)} = K_p + T_i s^{-\lambda} + T_d s^\delta$. This $PI^\lambda D^\delta$ controller with complex zeros and poles located anywhere in the left-hand side $s \rightarrow w$ plane, may be rewritten as:

$H(s) = K \frac{(s/\omega_n)^{\delta+\lambda} + (2\varsigma s^\lambda)/\omega_n + 1}{s^\lambda}$, where K is a gain, ς is the dimensionless damping ratio (is chosen mostly as under-damped, $\varsigma < 1$ and $= 1$ for critically damped), and ω_n is the natural frequency. This $PI^\lambda D^\delta$ is generalization to TID (Tilt Integral Derivative) compensator, which has a similar structure as PID, but the proportional component is replaced with tilted component having a transfer function s, to the power of $(-1/n)$. Therefore, the transfer function of TID is $H(s) = \frac{T}{s^{1/n}} + \frac{I}{s} + Ds$, where T, I, D are controller constants and n is the non-zero real number (between 2 and 3). The above transfer function approximates Bode's ideal transfer function (US patent 5-371-670).

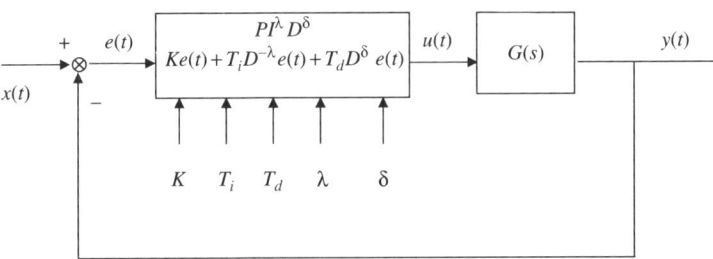

Fig. 9.3 Fractional order PID controls

Fig. 9.4 Fractional order PID and integer order PID, PD, PI

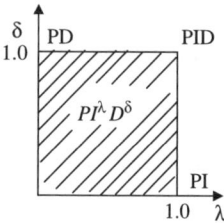

It can also be mentioned that the controller can also be characterized with fractional Laplace operator by band-limited lead effect. The lead lag compensator is of the representation $H(s) = K\left(\frac{1+\tau_1 s^q}{1+\tau_2 s^q}\right), \tau_2 < \tau_1$ and q being the fractional number. This transfer function can be realized by rational function approximation methods by recursive distribution of poles and zeros.

The example in state variable formulation gives some light about PID controls in fractional domain. Let a fractional plant transfer characteristics $G(s)$ be identified as

$$a_2 D^\alpha y(t) + a_1 D^\beta y(t) + a_0 y(t) = u(t)$$

simply written as $a_2 y^{(\alpha)}(t) + a_1 y^{(\beta)}(t) + a_0 y(t) = u(t)$. The fractional order PID $H(s)$ is

$$u(t) = Ke(t) + T_i D_t^{-\lambda} e(t) + T_d D_t^\delta e(t)$$

Using state variable $y(t) = x(t) = x_1(t)$ and $x^{(\beta)}(t) = x_2(t)$, the $G(s)$ is obtained as state-space, as:

$$\begin{bmatrix} x_1^{(\beta)}(t) \\ x_2^{(\alpha-\beta)}(t) \end{bmatrix} = \begin{bmatrix} 0 & 1 \\ -\frac{a_0}{a_2} & -\frac{a_1}{a_2} \end{bmatrix} \begin{bmatrix} x_1(t) \\ x_2(t) \end{bmatrix} + \begin{bmatrix} 0 \\ \frac{1}{a_2} \end{bmatrix} \begin{bmatrix} 0 \\ u(t) \end{bmatrix} \text{ and } y(t) = x_1(t)$$

By GL definition ${}_aD_t^{\pm r} f(t) = \lim_{T \to 0} \frac{1}{T^{\pm r}} \sum_{j=0}^{\left[\frac{t-a}{T}\right]} b_j^{\pm r} f(t-jT)$,

where $b_0^{\pm r} = 1, b_j^{\pm r} = \left(1 - \frac{(1+(\pm r))}{j}\right) b_{j-1}^{\pm r}$ are the binomial coefficients. The discretized state-space expression for $G(s)$ is

$$x_{1,k+1} = -\sum_{j=1}^{k+1} b_j^\beta x_{1,k+1-j} + T^\beta x_{2,k}$$

$$x_{2,k+1} = -\sum_{j=1}^{k+1} c_j^{\alpha-\beta} x_{2,k+1-j} + T^{\alpha-\beta} \left(-\frac{a_0}{a_2} x_{1,k} - \frac{a_1}{a_2} x_{2,k} + \frac{1}{a_2} u_k\right)$$

$$y_k = x_{1,k}$$

9.6 Feedback Control System

The transfer characteristic equation in the state-space format is discretized using the definition of Grunwald–Letnikov (GL) differintegral. For the state variables, fractional derivative $x_1^{(\beta)} = D_t^\beta x_1(t)$ and $x_2^{(\alpha-\beta)} = D_t^{(\alpha-\beta)} x_2(t)$, GL expansion is made and then written in the above format.

As we see fractional order differential equations accumulate the whole information of the formula in a weighted form, this is called the "memory effect." Fractionally differentiated state-space variable must be known as long as the system has been operated to obtain the correct response. This is known as "initialization function." For integer order systems, it is constant, and for fractional order systems, it is time varying. In the usual integer order system theory, the set of states of the system known at any given point in time along with the system equations are sufficient to predict the response of the system, both forward and backward in time. The fractional dynamic variables do not represent the "state" of the system at any given time in the previous sense, we need "history" of states or sufficient number by "short-memory principle" for initialization function computation. Because of the above-mentioned memory effect from the high-memory consumption, advance control algorithms with direct discretization of Tustin rule, Al-Alouni rule, by power series expansion (PSE) and/or continued fraction expansion (CFE) is developed. In these new approaches, the memory requirement is one tenth of the memory required for GL method. Several advance algorithms even to reduce this memory requirement is an interesting topic of research and development.

Many generalization of integer control design are possible with freedom allowed by fractional order systems. Some of them are listed below:

a. *Integral control*: Fractional integrals, $H(s) = ks^{-q}$, are used as compensators. The interesting feature of fractional integrals is that they still allow closed-loop tracking of step reference signals, while allowing the freedom to tune the low-frequency and high-frequency behavior by tuning the value of q, although the tracking will be slower.

b. *Derivative controls*: Although the pure derivative control is seldom used, derivatives of any fractional order, $H(s) = ks^q$, are available, and these will have less noise amplification at high frequencies than integer order derivatives.

c. *PI, PD, and PID controls*: Fractional elements allow use of any value of q, for integral and derivative in these controllers. If a fractional PID control is implemented, the fraction in the derivative part need not be same as the integral part fraction. The different fractions are indicated in Fig. 9.3, the controller will be $H(s) = k_p + k_i s^{-q_1} + k_d s^{q_2}$. The generalizations of the PID controls give a research topic of continuum order distribution controller (Chap. 9.9).

d. *Lead and lags*: Lead compensators are often used to help stabilize marginally stable system. Lag compensators are often used to reduce the magnitude of high-frequency loop gain of the system. Using fractional order components, it is possible to design fractional leads and fractional lag $H(s) = \frac{k(s^{q_1}+a)}{s^{q_2}+b}$. The benefit to these is that it is easier to shape the open-loop and closed-loop frequency responses using them than exclusively integer-order elements, due to the extra freedom offered by the continuum values of q.

e. *Start point singularity*: Design criterion unique to fractional order system deals with time domain singularity occurring at time zero. If the plant transfer function $G(s)$ does not contain a term in the denominator with an exponent of at least 1, that is the leading term in the denominator is s^q, then by the initial value theorem $g(0) = \lim_{s \to \infty} sG(s)$, the impulse response, $g(t) = L^{-1}\{G(s)\}$, will have a singularity at time zero. This may not be desirable, however using appropriate compensator in the forward path (say $H(s) = ks^{q-1}$), the singularity in the output of plant will be eliminated.

9.6.1 Concept of Iso-damping

The concept of having iso-damping i.e. overshoot independent of the payload (or system gain) has remarkable usage in the field of control sciences. This is only possible by use of fractional order calculus theory. This concept was introduced as an example of DC motor controls in Chap. 3. The plant may or may not be of fractional order, but controller with fractional order differential and or integral action is what makes the system response iso-damped.

This concept when applied to nuclear power plant controls elevates overall fuel efficiency robustness. The nuclear reactor is divided in two parts of, namely reactor core and coupled energy transfer equipments (heat exchangers, boiler, turbine generator). For the reactor part, the error correction is carried out in terms of ratio control of the observed neutron power (feedback) to the demanded power of the set point (on suitable set exponent). The effective power error is written as:
$$EPE = K_1(\log\{CalP\} - \log\{DemP\}) + K_2\left(\frac{1}{T_{OBS}} - \frac{1}{T_{SET}}\right) \propto \frac{CalP.e^{t/T_{OBS}}}{DemP.e^{t/T_{SET}}},$$
i.e. the ratio of observed neutronic power to the demanded power. This gives better results as compared to the existing integer order linear PID type corrections, as the governing formula is close to the reactor physics (which follows exponential and logarithmic expressions). In above expression $K_1 = 1$ and $K_2 = 0.5$ s (digitization time of control computer). This expression error governance gives fuel-efficient concept.

The reactor is coupled to several energy conversion devises, which governed by fractional order PID concept with iso-damping will increase overall nuclear power plant efficiency. The systems are always under dynamic corrections, and there are changes in gains (say payloads) and parametric shifts. This will be compensated by having iso-damped control systems where overshoot is independent of gain (payloads). The systems thus coupled to nuclear reactor will load the same with constant overshoot throughout, thus it will have the same energy transfer and not jittery. This process integrated over life span of nuclear power station will save nuclear fuel. This example to govern large energy transfer machines with fractional order controller with iso-damping is true for any power plant thus, to enhance overall fuel efficiency.

9.6 Feedback Control System

During 'Start-Up Operation' of nuclear plants, while negative feedback stabilizing factors are absent, is a risky affair. Risk factor is more for experimental research reactor, where reactor cores are configurable with several configurations and materials. Although reactor trip systems will take care of any excursions in power levels at start-up experiments, yet governing the start-up procedure with a fractional order feedback controller where overshoot is independent of gain (fuel characteristics) will add another level of safety and confidence.

As an example of DC motor controls, let the motor (plant) be expressed as $G(s) = \frac{K_m}{Js(s+1)}$ with J as payload inertia. Let the selected fractional open-loop transfer function have robustness and stability measure as per Bode's ideal function with phase margin constant of 60^0. Then Bode's ideal open-loop transfer function, which gives this phase margin, is $G_0(s) = \frac{1}{s^{4/3}}$. Since $G_0(s) = H(s)G(s)$, we can obtain a transfer function in the form $H(s) = \frac{J}{K_m}\left(s^{2/3} + \frac{1}{s^{1/3}}\right) = K\left(s^{2/3} + s^{-1/3}\right)$, which is a particular case of $PI^\lambda D^\delta$, where $K = J/K_m$ is the controller constant. The phase margin of the controlled system with a forward-loop controller is $\Phi_m = \arg[H(j\omega_0)G(j\omega_0)] + \pi = \arg\left[\frac{1}{(j\omega)^{4/3}}\right] + \pi = \pi - \frac{4}{3}\frac{\pi}{2} = \frac{\pi}{3}$, where ω_0 is cross-over frequency.

The constant phase margin is not dependent on payload (gain) changes and the system gains K. The phase curve is a horizontal line at $-2\pi/3$.

Step response of the closed loop can be expressed as:

$y(t) = L^{-1}\left\{\frac{1}{s(s^{1+1/3}+1)}\right\} = t^{1+1/3}E_{1+1/3,2+1/3}\left(-t^{1+1/3}\right)$, where step response is independent of the payload inertia at fractional order setting of 4/3. The solution is noted as follows.

Close-loop transfer function from open-loop transfer function $G_0(s) = s^{-4/3}$ is

$$G_{CL}(s) = \frac{G_0(s)}{1 + G_0(s)} = \frac{1}{s^{4/3} + 1}.$$

The step input will have output (time) response as:

$$y(t) = L^{-1}\left\{\frac{1}{s\left(s^{4/3}+1\right)}\right\} = L^{-1}\left\{\frac{1}{s}\left[1 - \frac{s^{4/3}}{s^{4/3}+1}\right]\right\}$$

$$= L^{-1}\left\{\frac{1}{s} - \frac{s^{\frac{4}{3}-1}}{s^{4/3}+1}\right\} = 1 - E_{4/3}(-t^{4/3})$$

Expanding with definition of one-parameter Mittag-Leffler function $E_{4/3}(-t^{4/3})$ we get

$$y(t) = 1 - \left[1 + \frac{(-t^{4/3})}{\Gamma(\frac{4}{3}+1)} + \frac{(-t^{4/3})^2}{\Gamma(2.\frac{4}{3}+1)} + \frac{(-t^{4/3})^3}{\Gamma(3.\frac{4}{3}+1)} + \cdots \right]$$

$$= -\frac{(-t^{4/3})}{\Gamma(\frac{4}{3}+1)} - \frac{(-t^{4/3})^2}{\Gamma(2.\frac{4}{3}+1)} - \frac{(-t^{4/3})^3}{\Gamma(3.\frac{4}{3}+1)} - \cdots$$

$$= \frac{t^{4/3}}{\Gamma(\frac{4}{3}+1)} - \frac{(t^{4/3})^2}{\Gamma(2.\frac{4}{3}+1)} + \frac{(t^{4/3})^3}{\Gamma(3.\frac{4}{3}+1)} - \cdots$$

$$= t^{4/3} \left[\frac{1}{\Gamma(\frac{4}{3}+1)} + \frac{(-t^{4/3})}{\Gamma(2.\frac{4}{3}+1)} + \frac{(-t^{4/3})^2}{\Gamma(3.\frac{4}{3}+1)} + \cdots \right]$$

Using the definition of two-parameter Mittag-Leffler function the following is obtained:

$$y(t) = t^{4/3} E_{\frac{4}{3},\frac{4}{3}+1}(-t^{4/3}) = t^{1+\frac{1}{3}} E_{1+\frac{1}{3},2+\frac{1}{3}}(-t^{1+\frac{1}{3}})$$

9.6.2 Fractional Vector Feedback Controller

This section considers the use of fractional vector feedback for control system design, especially for multivariate fractional state-space. The modeling of the system in fractional vector state-space has been covered in Chap. 7, and example is given in Chap. 8. The vector representation is

$_0 d_t^q \bar{x}(t) + \bar{\psi}(\bar{x}, q, a, 0, t) = A\bar{x}(t) + B\bar{u}(t)$ and $\bar{y}(t) = C\bar{x}(t) + D\bar{u}(t)$. Typically, vector feedback is implemented as $\bar{u}(t) = -K\bar{x}(t) + \bar{r}(t)$, where $\bar{r}$ is the vector set point and K is the feedback gain matrix to be determined. The closed-loop system then becomes

$$_0 d_t^q \bar{x}(t) = [A - BK]\bar{x}(t) - \bar{\psi}(\bar{x}, q, a, 0, t) + B\bar{r}(t)$$
$$\bar{y}(t) = [C - DK]\bar{x}(t) + D\bar{r}(t)$$

By choosing K appropriately and using standard pole-placement method, it is possible to place the eigenvalues anywhere in the w–plane. The det $[wI - A + BK]$ will determine the pole-placement. While doing this, the eigenvalues should be placed to the left of the instability wedge (Chap. 7). In this, $s^q = w$; it is the transformation from $s \to w$ plane. Presently without worrying about the instability due to initialization vector $\bar{\psi}$, the pole-placement can be carried to give the stable closed-loop performance as desired. At present, it is unclear how linear quadratic regulator (LQR) and other optimal feedback regulator rules are redrawn for fractional vector state problems.

9.6 Feedback Control System

Consider $A = \begin{bmatrix} 0 & 1 & 0 \\ 0 & 0 & 1 \\ -2 & -4 & -6 \end{bmatrix}$, $B = \begin{bmatrix} 0 \\ 0 \\ 1 \end{bmatrix}$, $C = \begin{bmatrix} 1 & 2 & 3 \end{bmatrix}$, $D = [0]$, the system is ${}_0 D_t^q \bar{x}(t) = A\bar{x} + B\bar{u}$ and $\bar{y}(t) = C\bar{x}(t)$ with controller as $\bar{u}(t) = -K\bar{x}(t)$. The control parameter (gains) will be chosen such that the poles in the w-plane are $w_{1,2} = -2 \pm j4$ and $w_3 = -6$. The Laplace transformation of the vector system equation with feedback regulator is $\bar{X}(s) = -[s^q I - A + BK]^{-1} \bar{\psi}(s)$. The close-loop system matrix is $[I s^q - A + BK] = [wI - A + BK]$, from here the value of K is evaluated as:

$$\det[wI - A + BK] = \det \left\{ \begin{bmatrix} w & 0 & 0 \\ 0 & w & 0 \\ 0 & 0 & w \end{bmatrix} - \begin{bmatrix} 0 & 1 & 0 \\ 0 & 0 & 1 \\ -2 & -4 & -6 \end{bmatrix} + \begin{bmatrix} 0 \\ 0 \\ 1 \end{bmatrix} \begin{bmatrix} k_1 & k_2 & k_3 \end{bmatrix} \right\}$$

$$\det \begin{bmatrix} w & -1 & 0 \\ 0 & w & -1 \\ 2+k_1 & 4+k_2 & w+6+k_3 \end{bmatrix} = w^3 + (6+k_3)w^2 + (4+k_2)w + (2+k_1)$$

From the poles given in the w-plane, the characteristic equation is.

$\alpha(w) = (w+2-j4)(w+2+j4)(w+6) = w^3 + 10w^2 + 44w + 120$. Comparing this with coefficients of the determinant, we get $K = \begin{bmatrix} k_1 & k_2 & k_3 \end{bmatrix} = \begin{bmatrix} 4 & 40 & 118 \end{bmatrix}$.

Here we have obtained the controller matrix and the gains because the system is controllable as rank of matrix $\begin{bmatrix} B & AB & A^2 B \end{bmatrix}$ is 3 full ranks. These are usual theory of multivariate control science. Refer Fig. 9.5 for fractional vector state feedback block diagram.

9.6.3 Observer in Fractional Vector System

Just as in integer order theory, it is important to create observers or vector estimators for fractional order system. This section will present the theory necessary for designing fractional order observer. The fractional order vector estimator is

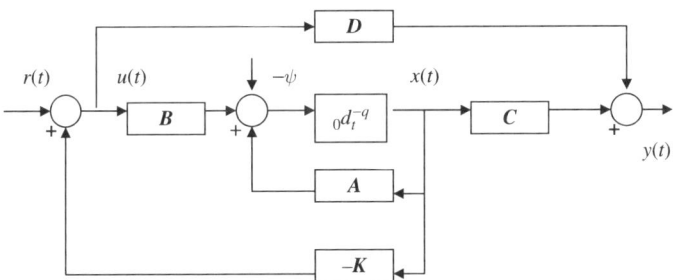

Fig. 9.5 State variable fractional vector controller

$$_0d_t^q \hat{\underline{x}}(t) + \hat{\underline{\psi}}(\underline{x}, q, a, 0, t) = A\hat{\underline{x}}(t) + B\underline{u}(t) - L\left[\underline{y}(t) - \hat{\underline{y}}(t)\right]$$

$\hat{\underline{y}}(t) = C\hat{\underline{x}}(t) + D\underline{u}(t)$, where a non-zero initialization function $\hat{\underline{\psi}}$ has been assumed for the observer. The vector error $\underline{e}(t)$ is defined as the difference between the real system outputs $\underline{x}(t)$ and the estimated observer outputs $\hat{\underline{x}}(t)$:

$\underline{e}(t) = \underline{x}(t) - \hat{\underline{x}}(t)$. The observer gain L is determined as to force the error between the plant vectors to go to zero. The dynamics of the error are obtained by fractionally differentiating the error equation as: $_0d_t^q \underline{e}(t) = {_0d_t^q} \underline{x}(t) - {_0d_t^q} \hat{\underline{x}}(t)$. In this, substitute the system equations to get

$$_0d_t^q \underline{e}(t) = \left[A\underline{x}(t) + B\underline{u}(t) - \underline{\psi}(\underline{x}, q, a, 0, t)\right]$$
$$- \left[A\hat{\underline{x}}(t) + B\underline{u}(t) - \hat{\underline{\psi}}(\hat{\underline{x}}, q, a, 0, t)\right] - L\left[\underline{y}(t) - \hat{\underline{y}}(t)\right]$$

Now replacing the sensed system outputs $\underline{y}(t)$ and $\hat{\underline{y}}(t)$ with the vector variable $C\underline{x}(t) + D\underline{u}(t)$ we obtain

$$_0d_t^q \underline{e}(t) = \left[A\underline{x}(t) + B\underline{u}(t) - \underline{\psi}(\underline{x}, q, a, 0, t)\right]$$
$$- \left[A\hat{\underline{x}}(t) + B\underline{u}(t) - \hat{\underline{\psi}}(\hat{\underline{x}}, q, a, 0, t)\right]$$
$$- L\left[C\underline{x}(t) + D\underline{u}(t) - C\hat{\underline{x}}(t) - D\underline{u}(t)\right]$$

Eliminating and doing simple algebra and substituting $\underline{x}(t) - \hat{\underline{x}}(t) = \underline{e}(t)$ we get

$$_0d_t^q \underline{e}(t) = [A - LC]\underline{e}(t) - \left\{\underline{\psi}(\underline{x}, q, a, 0, t) - \hat{\underline{\psi}}(\hat{\underline{x}}, q, a, 0, t)\right\}$$

The matrix L is determined to force the observer error to go to zero by placing the eigenvalues of $[A - LC]$ in the stable region of w-plane (Chap. 7). The initialization function response eventually decays to zero for any system with $0 < q < 1$ and only has transient effect on the observer error. However, proper choice of $\hat{\underline{\psi}}$ will help drive the error to zero sooner than if $\hat{\underline{\psi}}$ was simply zero.

Consider $_0D_t^q \underline{x}(t) = A\underline{x}(t) + B\underline{u}(t)$ and $\underline{y}(t) = C\underline{x}(t)$, observer gain has to set such that the closed-loop poles follow the w-plane as $w_{1,2} = -2 \pm j\sqrt{12}$, $w_3 = -5$.

9.6 Feedback Control System

$$A = \begin{bmatrix} 0 & 1 & 0 \\ 0 & 0 & 1 \\ -6 & -11 & -6 \end{bmatrix}, B = \begin{bmatrix} 0 \\ 0 \\ 1 \end{bmatrix}, C = \begin{bmatrix} 1 & 0 & 0 \end{bmatrix}, D = [0]$$

The normal observability matrix is $\begin{bmatrix} C \\ CA \\ CA^2 \end{bmatrix} = \begin{bmatrix} 1 & 0 & 0 \\ 0 & 1 & 0 \\ 0 & 0 & 1 \end{bmatrix}$ and has full rank. Therefore, the system is observable. The characteristic polynomial is

$$\alpha(w) = (w + 2 - j\sqrt{12})(w + 2 + j\sqrt{12})(w + 5) = w^3 + 9w^2 + 36w + 80.$$

$$\det[wI - A + LC] = \det \left\{ \begin{bmatrix} w & 0 & 0 \\ 0 & w & 0 \\ 0 & 0 & w \end{bmatrix} - \begin{bmatrix} 0 & 1 & 0 \\ 0 & 0 & 1 \\ -6 & -11 & -6 \end{bmatrix} + \begin{bmatrix} l_1 \\ l_2 \\ l_3 \end{bmatrix} \begin{bmatrix} 1 & 0 & 0 \end{bmatrix} \right\}$$

$$\det \begin{bmatrix} w + l_1 & -1 & 0 \\ l_2 & w & -1 \\ l_3 + 6 & 11 & w + 6 \end{bmatrix} = w^3 + (l_1 + 6)w^2 + (6l_1 + l_2 + 11)w$$
$$+ (11l_1 + 6l_2 + l_3 + 6)$$

Equating the determinant to the characteristic polynomial, we obtain $l_1 + 6 = 9$, $6l_1 + l_2 + 11 = 36$, $11l_1 + 6l_2 + l_3 + 6 = 80$ which give the value for observer gain as $L = \begin{bmatrix} l_1 & l_2 & l_3 \end{bmatrix} = \begin{bmatrix} 3 & 7 & -1 \end{bmatrix}$.

Refer Fig. 9.6 for fractional vector observer block diagram.

9.6.4 Modern Aspects of Fractional Control

The multivariate control aspects are discussed in the previous sections with examples. The controllable and observable issues are discussed, and disregarding the

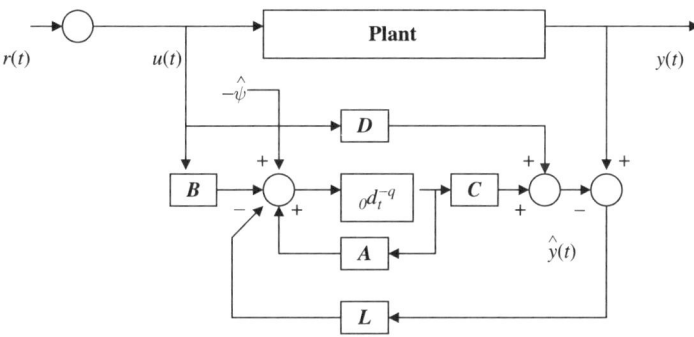

Fig. 9.6 Fractional vector dynamic variable observer

time varying initialization vector, the gains of fractional state feedback and gains of observer were calculated. In the example, coefficient equalization scheme is chosen to equalize the determinants of controller and observer matrices. Here the Bass-Gura and Ackerman formula from the rank determination approach is also suitable, to place the poles as desired in the w-plane stability wedge.

Question of controllability and observability and minimality arise when the systems are expressed in the vector fractional dynamic variable form. There are two directions to go in this regard. One is to completely redefine and derive all of the related vector system properties. This is not done at this time though effort is on in this regard. It is important to note that the fractional dynamic variable vector alone does not contain all the information about the state of the system, but requires vector $\bar{\psi}$, the initialization vector. The implication of this is totally not clear at this stage; however, one important consideration in observability and controllability issue is the inclusion of the time-dependent initialization function. Rather than re-deriving all the system theory results on controllability and observability, useful results can still be obtained by simply using the observabilty and controllability matrices directly for particular fractional vector system—while neglecting the initialization vector. This can be done on usual *ABCD* matrices completely without regard to their deeper theoretic implementation.

Finally, it should be noted that the vector space of fractional dynamic variables allows direct use of standard state variable feedback and observer theory, with the understanding that the closed-loop poles are being placed in stability zone of w-plane. It is not clear at this point how to interpret any optimal control theory rules. Although we could use Lyapunov and Riccati equations for design, their interpretation is not clear with regard to optimality. One would expect that the resulting controllers, which always guarantee to have closed-loop poles in left-half plane (LHP) of s-domain, would now place all closed-loop poles in the LHP of w-domain, which would guarantee some form of stable "hyper-damped" response.

Kalman's decomposition laws are still valid with initialization function taken as one of the plant disturbance input. Also for state estimator, "fractional Kalman filter" and extended, "fractional Kalman filter" for linear and non-linear systems are recent developmental fields of multivariate control science.

9.7 Viscoelasticity (Stress–Strain)

Stress relaxation and creep behavior in stress–strain relationship are well described by fractional order models. A stress–strain law for viscoelastic materials is described as $\varepsilon(t) = \frac{1}{K} {}_c D_t^{-v} \sigma(t)$, or a new specimen, where no initialization is required and it is thus memoryless which is represented as (un-initialized derivative) $\varepsilon(t) = \frac{1}{K} {}_a d_t^{-v} \sigma(t)$. Here ε is the strain and σ the is stress. For $v = 0$, the material is elastic solid and for $v = 1$ the material is viscous liquid. K is constant which depends on the material. In the above stress–strain relationship both effects, i.e., instantaneous elastic and long-term viscous flow, are neglected.

9.7 Viscoelasticity (Stress–Strain)

Consider a unit step load is applied from $t = 0$ and $t = d$ on a new specimen (un-initialized). $\sigma(t) = \kappa [H(t) - H(t - d)]$, H is Heaviside step unity function and κ is the magnitude. Then:

$$\varepsilon(t) = \frac{\kappa}{K\Gamma(v)} \int_0^t (t - \tau)^{v-1} [H(\tau) - H(\tau - d)] d\tau$$

$$= \frac{\kappa}{K\Gamma(v)} \left[\int_0^t (t - \tau)^{v-1} d\tau H(t) - \int_0^t (t - \tau)^{v-1} d\tau H(t - d) \right]$$

$$= \frac{\kappa}{K\Gamma(v+1)} \left[t^v H(t) - (t - d)^v H(t - d) \right]$$

This defines the strain (creep) response, for the prescribed loading. Figure 9.7 gives the strain curve for this prescribed loading. For $t < d = 1$, the curve is essentially a creep function, and for $t > d = 1$, is it a relaxation period.

The above example material was new. For creep initialization function associated with constant past loading may be readily inferred from

$$\varepsilon(t) = \frac{\kappa}{K\Gamma(v+1)} \left[t^v H(t) - (t - d)^v H(t - d) \right]$$

as follows.

Consider this problem initialized at point c, then $_c D_t^{-q} f(t) = {_a D_t^{-q}}$, for $t > c$; therefore, $_c d_t^{-q} f(t) + \psi(f, -q, a, c, t) = {_a D_t^{-q}} f(t) = {_a d_t^{-q}} f(t)$, for $t > c$. Thus ψ may be expressed as $\psi(f, -q, a, c, t) = {_a d_t^{-q}} - {_c d_t^{-q}}$, for $t > c$.

For obtaining the initialization function of creep problem initialized at $t = c = d$, where d is as used in the above equation, i.e.,

$$\varepsilon(t) = \frac{\kappa}{K\Gamma(v+1)} \left[t^v H(t) - (t - d)^v H(t - d) \right].$$

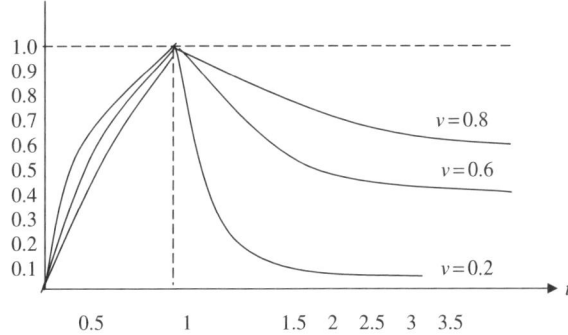

Fig. 9.7 Strain versus time (creep function and relaxation function)

Applying this result to equation $\psi(f, -q, a, c, t) = {}_a d_t^{-q} - {}_c d_t^{-q}$, to the creep problem, gives

$$_c D_t^{-v} \sigma(t) = {}_c d_t^{-v} \sigma(t) + \psi(\sigma, -v, a, c, t), \text{ for } t > c$$

Then taking $c = d$ and since $\sigma(t) = 0$ for $t > d$, we get

$$_c d_t^{-v} \sigma(t) = {}_a d_t^{-v} \sigma(t) = 0, \text{ for } t > c = d,$$

and K times the response equation of

$$\varepsilon(t) = \frac{K}{K\Gamma(v+1)} \left[t^v H(t) - (t-d)^v H(t-d) \right]$$

is the initialization function $\psi(t)$; therefore:

$$\psi(\sigma, -v, a, c, t) = \frac{K}{\Gamma(v+1)} \left[t^v H(t) - (t-c)^v H(t-c) \right], \text{ for } t > c = d.$$

This initialization function with proper time shifts may thus be employed for the material, which had "creep history."

The stress relaxation (or creep) function $\chi(t)$ is defined as the stress required to produce a strain $H(t)$, excluding terms that are initially infinite or do not tend to zero as time grows. For a new material this stress is

$$\chi(t) = \sigma(t) = K {}_a d_t^v \varepsilon(t) = K {}_a d_t^1 {}_a d_t^{-(1-v)} \varepsilon(t)$$

$$= K \frac{d}{dt} \frac{K}{\Gamma(1-v)} \int_0^t (t-\tau)^{-v} H(\tau) d\tau \chi(t) = \frac{K}{\Gamma(1-v)} t^{-v} H(t)$$

yielding as a result the stress (creep) relaxation function for this formulation.

This stress–strain with memory and history definitions with fractional calculus is a suitable method of describing the fundamentals of "shape-memory" alloys.

9.8 Vibration Damping System

The addition of damping is important in many applications for stability enhancement. One such is requirement to have vibration-damping augmentation in gas turbines blades (specially for aerospace). Figure 9.8 gives spring mass viscodamped dynamic system diagram, and we derive the transfer function of the same.

Since the transfer function requires the initialization to be zero, the equations describing the system use un-initialized operators as follows:

9.8 Vibration Damping System

Fig. 9.8 Spring mass viscodamped dynamics

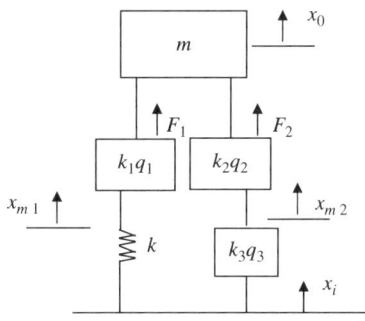

$$F_1 = -k_{10}d_t^{q_1}(x_0 - x_{m1})$$
$$F_1 = k(x_{m1} - x_i)$$
$$F_2 = -k_0 d_t^{q_2}(x_0 - x_{m2})$$
$$F_2 = k_{30}d_t^{q_3}(x_{m2} - x_i), \text{ and}$$
$$F_1 + F_2 = m_0 d_t^2 x_0$$

where ks are the damping coefficients and the spring constants, and m, F, and x are the mass, force, and displacement respectively.

After taking Laplace transforms and considerable algebra and elimination process, the generalized transfer function is

$$\frac{X_0(s)}{X_i(s)} = \frac{\frac{1}{k_2}s^{q_1+q_3} + \frac{1}{k_3}s^{q_1+q_2} + \frac{1}{k_1}s^{q_2+q_3} + \frac{1}{k}s^{q_1+q_2+q_3}}{\frac{m}{k_1 k_2}s^{2+q_3} + \frac{m}{kk_2}s^{2+q_1+q_3} + \frac{m}{k_1 k_3}s^{2+q_2} + \frac{m}{kk_3}s^{2+q_1+q_2} + \frac{1}{k_2}s^{q_1+q_3} + \frac{1}{k_3}s^{q_1+q_2} + \frac{1}{k_1}s^{q_2+q_3} + \frac{1}{k}s^{q_1+q_2+q_3}}$$

This general transfer function can be studied for special cases, by appropriate choices of k, k_i, and q_i. For example, to allow damper 3 to represent a conventional dashpot-damper, let $q3 = 1$, and select appropriate k_3. Now making the $q1 = q2 = 0$ means eliminating the dampers and spring of infinite stiffness, i.e., $k_1 = k_2 = \infty$ gives simple transfer function of fractional order $q3$ with viscoelastic element 3. The transfer function is

$$\frac{X_0(s)}{X_i(s)} = \frac{\frac{k_3}{m}s^{q_3} + \frac{k}{m}}{s^2 + \frac{k_3}{m}s^{q_3} + \frac{k}{m}}$$

The frequency response may now be evaluated by putting $s \to j\omega$, and determining the magnitude and phase angle using the derived transfer function, for various values of k/m, k_3/m, and letting the fractional order $q3$ vary from 0 to 2 in steps of 0.2, to arrive at designated design measures.

In nature, we always get mix of mass spring and damper systems, and thus to have reality modeling, this fractional calculus plays an important role.

9.9 Concluding Comments

As far the imagination goes, the reality systems have the effect of past. The present behavior is someway definitely related to past history. The actual systems therefore are non-Markovian systems with past history, heredity, and memory. The fractional calculus thus is a language what nature understands the best. This way, system modeling and control becomes efficient. The application here included simple aspects of system modeling and control, and further research is required to have optimal descriptor of dynamic vector space controllers. Not only for science and engineering this tool is an efficient descriptor, but also for finance and stock market analysis this tool is being recently explored and applied. Therefore, this 300-year-old (new) subject will revive in this century, to speak what nature understands the best.

Chapter 10
System Order Identification and Control

10.1 Introduction

For unknown systems, "system identification" has become the standard tool of the control engineer and scientists. Identifying a given system from data becomes more difficult, however when fractional orders are allowed. Here identification process is demonstrated using the assumption that system order distribution is a continuous one. Frequency domain system identification can thus be performed using numerical methods demonstrated in this chapter. Here one concept of rLaplace transforms is discussed (Laplace transform in log domain), to discuss the system order distribution. Also mentioned is the variable order identification as further development, where the system order also varies with ambience and time is highlighted. Here in this chapter, an identification method based on continuous order distribution is discussed. This technique is suitable for both the standard integer order and fractional order systems. This is topic of further advance research, as to qualify the procedure of system order identification and to have a technique for tackling variable order. Extending this continuous order distribution discussion, the advance research of having a continuum order feedback and generalized PID control is elucidated. Also in this chapter some peculiarities of the pole property of fractional order system as ultra-damping, hyper-damping, and fractional resonance are explained. Elaborate research in this direction is an ongoing process; to crisply define the system identification and the variable order structure, along with generalized controller for future applications.

10.2 Fractional Order Systems

As the concept of order is central to the understanding of fractional (or integer) order systems, some discussion of this concept now follows. In this discussion, single-input–output systems are considered. The examples in the earlier chapters for heat flow and transmission line (lossy and lossless), and several examples gave the stage for half-order system or zero-order system, integer order systems, and fractional order systems. Recalling the characteristic equations or transfer function definition, we call a system first order, second order, third order, etc., similarly, the system can be of fractional order too, i.e., the characteristic equation having powers of s-variable of non-integers numbers (real). We also consider that system representation is generally of "minimal phase," and they are linear. For non-minimal phase, the system behavior is for a positive step demand; the output initially goes in reverse and then changes direction to follow the demand. This is peculiar of gas turbines, where for non-minimal phase system, the velocity first will marginally decrease and then eventually increase to the follow increment in the demanded velocity.

Mathematical order conventionally is defined as the highest derivative occurring in a given differential equation. The concept of mathematical order is applicable to both ordinary and fractional differential equations. Normally, when the order is used without qualifier, it implies the meaning of mathematical order.

For linear dynamic systems that are described by ordinary differential equations, the system mathematical order implies or is equivalent to the following:

(1) The highest derivative in ordinary differential equation
(2) The highest power of Laplace variable-s, in the characteristic equation
(3) The number of initializing constants required for the differential equation
(4) The length of the state vector
(5) The number of singularities in the characteristic equation
(6) The number of energy storage elements
(7) The number of independent spatial directions in which a trajectory can move
(8) The number of devices that can add 90° sinusoidal steady-state phase lag, and
(9) The number of devices that retain some memory of the past.

The utility of the definition of mathematical order for a differential equation, composed of integer order component, is that, to infer the characteristic behaviour and response of a system.

Thus the benefit of having a definition for order for linear ordinary differential equations is that it allows a direct understanding of the behavior of a given dynamic system. Unfortunately, for fractional differential equations, the order of the highest derivative does not infer all of the previously mentioned properties. Indeed, the most important characteristics of order in integer order ordinary differential equation is probably item (3) i.e. it indicates the number of initializing constants, which together with the differential equations allow prediction of the future behavior. In system terminology, this information provides initial "states," of the system being analyzed. Clearly, the order of highest derivative in a fractional differential equation does not have this property nor does it predict the associated number of energy/memory elements associated with fractional differential equation, nor does it infer the number of integrations (even fractional), required to solve simulate the given fractional differential equation. Thus the issue of order and the information required together with the fractional differential equation to predict the future is fundamental and should be treated differently. This is explained in section 10.12, that a seemingly first order characteristic polynomial with fractional order terms may go into resonance.

10.3 Continuous Order Distribution

For integer order systems once the maximum order of the system to be identified is chosen, the parameters of the model can be optimized directly. For fractional order systems, the identification requires (a) the choice of the number of fractional operators, (b) the fractional powers of the operator, and (c) the coefficients of the operators. Specifically in electrode–electrolyte interface experiments for determining the interface impedances, the frequency domain techniques are followed, for chosen transfer function. But to identify the form of transfer function itself, with order and coefficient, an approach from experimental data of frequency domain analysis should be the starting point, to identify unknown systems.

10.3 Continuous Order Distribution

A very basic of mass spring damper system of force balance is taken here to study the concept of continuous order distribution. The familiar system is represented (with un-initialized derivative) as:

$$m_0 d_t^2 x(t) + b_0 d_t^1 x(t) + kx(t) = f(t) \tag{10.1}$$

where $x(t)$ is the position of the mass m, $f(t)$ is the forcing function on the mass, b is the damping, and k is the spring (restoring) force. In the Laplace domain, this takes the following form:

$$(ms^2 + bs + k)X(s) = F(s) \tag{10.2}$$

It is well known that some element intermediate between spring and dashpot behaves and balances the forces called viscoelastic element. Such element is described as:

$$k_{q0} d_t^q x(t) = f(t) \qquad 0 \le q \le 1 \tag{10.3}$$

The Laplace representation is

$$k_q s^q X(s) = F(s) \tag{10.4}$$

Adding this viscoelastic element to the original assumed (lumped) system (10.2), we get

$$\left(ms^2 + bs + k_q s^q + k\right) X(s) = F(s) \tag{10.5}$$

It is known that viscoelastic elements will posses any order q between 0 and 1, so we can add another viscoelastic element and then keep on adding several others too, like the following:

$$\left(ms^2 + bs + k_{q2}s^{q2} + k_{q1}s^{q1} + k\right) X(s) = F(s) \qquad (10.6)$$

$$\left(ms^2 + k_{q4}s^{q4} + k_{q3}s^{q3} + bs + k_{q2}s^{q2} + k_{q1}s^{q1} + k\right) X(s) = F(s) \qquad (10.7)$$

This process could be continued so that the system can therefore be expressed as power series with $0 \leq q_n \leq 2$, with N as integer, as follows:

$$\left(\sum_{n=0}^{N} k_n s^{q_n}\right) X(s) = F(s) \qquad (10.8)$$

Now in reality the order q_n is temperature dependent, and the entire system will be considered as layered of such material. Therefore if the material is subjected to temperature distribution, then the material will exhibit order distribution too. In the limit of infinitesimally small elements, the above (10.8) will tend to continuum, and the summation will be replaced then by integral. This gives the fundamental motivating procedure for the concept with continuous order distribution. This continuous order is expressed as:

$$\left(\int_0^2 k(q) s^q dq\right) X(s) = F(s) \qquad (10.9)$$

This is very general representation of a dynamic system of any type taken for system identification studies. For demonstration, the familiar integer order dynamic spring damper mass element equation (10.2) can be rewritten with the form expressed in (10.9) as:

$$\left(\int_0^2 [m\delta(q-2) + b\delta(q-1) + k\delta(q)] s^q dq\right) X(s) = F(s) \qquad (10.10)$$

Figure 10.1 shows the plot of $k(q)$ and q, for the classical (10.2) mass–spring–damper, i.e. order distribution. Here the order is discrete, dirac-delta functions at 0,1,2.

Figure 10.1 demonstrates for ideal (classical) systems; the orders are concentrated at a single number, in this case at 0,1,2. This is accepted if the mass, spring, and damper are (really) ideal elements. In the classical calculus terminology, these

Fig. 10.1 Order distribution of mass–spring–damper integer order system

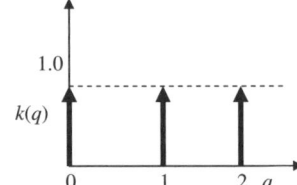

10.4 Determination of Order Distribution from Frequency Domain Experimental Data

orders are point property, the question is if at all these are point quantities? If so then should we have mass assigned to a point? But in reality, the mass is assigned a distributed volume and thus the idealism vanishes. Therefore, the order indeed need not be appoint property. Well even for integer order systems, these dirac-deltas concentrated at 0,1,2 should spread out may be like Gaussian distribution with peaks at 0,1,2 and spread from $q = 0$ to $+\infty$, in continuous spectrum. Though Fig. 10.1 shows the height of these dirac-delta functions to be same, but actually, the height is proportional to the coefficients. As per (10.10), the height will depend on the relative values of the multiplier coefficients m, b, k. To demonstrate the concept, all dirac-delta at the discrete order points are taken as of unity height, actually the heights should be different.

Allowed the restriction on the maximum possible order (i.e. second order in the present case), (10.9) can be still generalized as continuum power series, as following:

$$\left(\int_0^\infty k(q)s^q dq\right) X(s) = F(s) \qquad (10.11)$$

The time domain representation of the (10.11) is

$$\int_0^\infty k(q)_0 d_t^q x(t) dq = f(t) \qquad (10.12)$$

Mathematically, the system can also be described by continuum asymptotic series, instead of power series (10.11), as:

$$\left(\int_0^\infty k(q)s^{-q} dq\right) X(s) = F(s) \qquad (10.13)$$

Combining (10.11) and (10.13), we obtain a very general system descriptive method, as:

$$\left(\int_{-\infty}^{+\infty} k(q)s^q dq\right) X(s) = F(s) \qquad (10.14)$$

10.4 Determination of Order Distribution from Frequency Domain Experimental Data

Take the general power series representation of a system (10.11)

$$\left(\int_0^\infty k(q)s^q dq\right) X(s) = F(s)$$

It is desirable to get the transfer function:

$$\frac{X(s)}{F(s)} = \frac{1}{\int_0^\infty k(q)s^q dq} = \frac{1}{P(s)} = G(s) \tag{10.15}$$

Inverting (10.15) we get

$$P(s) = \int_0^\infty k(q)s^q dq = \frac{1}{G(s)} \tag{10.16}$$

For an unknown system, the measured frequency response is available experimentally, as $G(j\omega)$. In (10.16), we replace $s \to j\omega$, and assume that the representation of the system is by form (10.16). Then

$$\int_0^\infty k(q)(j\omega)^q dq = \frac{1}{G(j\omega)} \tag{10.17}$$

System identification problem thus boils down to finding the order distribution $k(q)$, given $G(j\omega)$.

Obviously, analytical approach is difficult, and we resort to numerical approach. In reality, the order distribution decays as $q \to \infty$. So to assume $k(q) \to 0$ as q grows will not be an offset from reality. In that event, the integral expressed in (10.17) converges and Euler's formula to numerically solve (10.17) is used. The integral of (10.17) will take the summation form as:

$$\sum_{n=0}^N k_n (j\omega)^{nQ} Q = \frac{1}{G(j\omega)} \tag{10.18}$$

where Q is the constant sample width, in the variable q, and k_n is the height of the sampled order distribution. Remembering that we usually have sampled frequency response data, (10.18) should be satisfied at each data point or for any frequency (ω_i). Thus we write (10.18) as:

$$\sum_{n=0}^N k_n (j\omega_i)^{nQ} Q = \frac{1}{G(j\omega_i)}, \text{ for any } i \tag{10.19}$$

Equation (10.19) can be written for 1 to M, frequency points as:

10.5 Analysis of Continuous Order Distribution

$$Qk_0 + Qk_1(j\omega_1)^Q + Qk_2(j\omega_1)^{2Q} + Qk_3(j\omega_1)^{3Q} + \ldots + Qk_N(j\omega_1)^{NQ} = \frac{1}{G(j\omega_1)}$$

$$Qk_0 + Qk_1(j\omega_2)^Q + Qk_2(j\omega_2)^{2Q} + Qk_3(j\omega_2)^{3Q} + \ldots + Qk_N(j\omega_2)^{NQ} = \frac{1}{G(j\omega_2)}$$

....
....

$$Qk_0 + Qk_1(j\omega_M)^Q + Qk_2(j\omega_M)^{2Q} + Qk(j\omega_M)^{3Q} + \ldots + Qk_N(j\omega_M)^{NQ} = \frac{1}{G(j\omega_M)} \quad (10.20)$$

Rearranging (10.20) in matrix form, we obtain

$$Q \begin{bmatrix} 1 & (j\omega_1)^Q & (j\omega_1)^{2Q} & . & (j\omega_1)^{NQ} \\ 1 & (j\omega_2)^Q & (j\omega_2)^{2Q} & . & (j\omega_2)^{NQ} \\ . & . & . & . & . \\ . & . & . & . & . \\ 1 & (j\omega_M)^Q & (j\omega_M)^{2Q} & . & (j\omega_M)^{NQ} \end{bmatrix} \begin{bmatrix} k_0 \\ k_1 \\ . \\ . \\ k_N \end{bmatrix} = \begin{bmatrix} 1/G(j\omega_1) \\ 1/G(j\omega_2) \\ . \\ . \\ 1/G(j\omega_M) \end{bmatrix} \quad (10.21)$$

The compact form representation is shown below:

$$[W] \cdot [k] = [g] \quad (10.22)$$

The $[W]$ matrix includes Q, and the vectors are $[k]$ and $[g]$. Clearly if $M > N$, then least square solution or matrix inverse solution may be applied, and can be solved to get sampled order distribution $k(q)$. The problem with the solution of (10.22) is that the matrix $[W]$ tends toward singularity if N (the number of order samples) grows large, or the order distribution sample size Q is made smaller.

Therefore numerically the order distribution $k(q)$ is $[k] = [W]^{-1}[g]$ or can be written as a pseudo-inverse expression, as $[k] = \left[[W]^T[W]\right]^{-1}[W]^T[g]$. The proven concepts of numerical robustness is maintained in these calculations too.

10.5 Analysis of Continuous Order Distribution

By rewriting the integral in system equation (10.11), i.e., $\int_0^\infty k(q)s^q dq$, the exponent $s^q = \exp[q \ln(s)]$, we obtain

$$\left(\int_0^\infty k(q)e^{q \ln(s)} dq\right) X(s) = P(s)X(s) = F(s) \quad (10.23)$$

The expression (10.23) is effectively a Laplace transform of the function $k(q)$, with the new Laplace variable $r = -\ln(s)$. As long as the order distribution $k(q)$ is of exponential order then the resulting $P(s)$ is easy to evaluate, using this r-Laplace transform.

Table 10.1 presents the transfer function $P(s)$, for second-order systems with continuous order distribution.

Equation (10.23) resembles Laplace transform in log frequency and could be used for frequency domain analysis by replacing $s \to j\omega_m$. Then

$$\int_0^\infty k(nQ)e^{nQ\ln(j\omega_m)} = P(j\omega_m),$$

rather difficult to solve, but solvable to obtain $k(nQ)$.

The order distribution $k(q)$ is taken for all $q \geq 0$. The system descriptions with the characteristic equations are expressed in differential equations with differentiation order greater than zero. So the integration terms are also converted to differentiation and the characteristic equations are in polynomial of powers of s^q with $q \geq 0$. In the following derivations thus, the q is always taken as greater than zero, and $k(q) = 0$, for all $q < 0$.

To evaluate r-Laplace transform from the given order distribution (continuous spiked or mixed), the Laplace identities are used. So $P(r) = \int_0^\infty k(q)e^{-rq}dq$ obtained is r-Laplace transform, and then in the obtained expression of $P(r)$ substitution for $r = -\ln(s)$ is carried out to get $P(s)$.

Following examples demonstrate the derivation of r-Laplace, $P(r)$, and then the Laplace transform $P(s)$ of continuous order, spiked ordered, and mixed order distributions, $k(q)$.

For exponential order distribution $k(q) = e^{-q}$, for all $q \geq 0$.
r-Laplace transform is

$$P(r) = \int_0^\infty e^{-q}e^{-rq}dq = \int_0^\infty e^{-q(r+1)}dq = \frac{1}{r+1},$$

in this expression, putting $r = -\ln(s)$ gives

$$P(s) = \frac{1}{1-\ln(s)}$$

For $k(q) = \frac{1}{q+a}$ for all $q \geq 0$,

$$P(r) = \int_0^\infty \frac{1}{q+a}e^{-qr}dq.$$

This integral we solve by using the definition of exponential integral, as:

10.5 Analysis of Continuous Order Distribution

Table 10.1 Order distributions and their (r - Domain and s - Domain) Laplace transform

order distribution $k(q)$ vs. q	r-Laplace transform $P(r)$	Laplace transfer function $P(s)$ $r = -\ln(s)$
	$\dfrac{1 - e^{-2r}}{r}$	$\dfrac{s^2 - 1}{\ln(s)}$
	$\dfrac{1 - 2e^{-r} + e^{-2r}}{r^2}$	$\dfrac{1 - 2s + s^2}{[\ln(s)]^2}$
	$\dfrac{1 - (1 + 2r)e^{-2r}}{r^2}$	$\dfrac{1 - [1 - 2\ln(s)]\,s^2}{[\ln(s)]^2}$
	$\dfrac{2r - 1 + e^{-2r}}{r^2}$	$\dfrac{s^2 - 1 - 2\ln(s)}{[\ln(s)]^2}$
	$\dfrac{re^{-2r} + 1 - e^{-r}}{r}$	$\dfrac{s - 1 + s^2 \ln(s)}{\ln(s)}$
	$\dfrac{r - 1 + e^{-r} + r^2 e^{-2r}}{r^2}$	$\dfrac{s^2[\ln(s)]^2 - \ln(s) + s - 1}{[\ln(s)]^2}$
	$\dfrac{1 - (1 + r)e^{-r} + r^2 e^{-2r}}{r^2}$	$\dfrac{1 - s + s\ln(s) + s^2[\ln(s)]^2}{[\ln(s)]^2}$
	$\dfrac{4\pi^2(1 - e^{-2r})}{2r(r^2 + 4\pi^2)}$	$\dfrac{4\pi^2(s^2 - 1)}{2\ln(s)\left[\{\ln(s)\}^2 + 4\pi^2\right]}$
	$\dfrac{1 - e^{-r}}{r}$	$\dfrac{s - 1}{\ln(s)}$
	$\dfrac{1 - (1 + r)e^{-r}}{r^2}$	$\dfrac{1 - s[1 - \ln(s)]}{[\ln(s)]^2}$

$$Ei(x) \stackrel{def}{=} -\int_{-x}^{\infty} \frac{e^{-t}}{t} dt$$

Put $u = q + a$, then

$$P(r) = \int_a^{\infty} \frac{1}{u} e^{-r(u-a)} du = e^{ar} \int_a^{\infty} \frac{1}{u} e^{-ru} du.$$

In this, take $ru = v$, then

$$P(r) = e^{ar} \int_{ar}^{\infty} \frac{r}{v} e^{-v} \frac{dv}{r} = e^{ar} \int_{-(-ar)}^{\infty} \frac{1}{v} e^{-v} dv = -e^{ar} \left\{ -\int_{-(-ar)}^{\infty} \frac{e^{-v}}{v} dv \right\}$$

$$= -e^{ar} Ei(-ar)$$

Substitute $r = -\ln(s)$ to get

$$P(s) = -e^{a[-\ln(s)]} Ei\left(-a\left[-\ln(s)\right]\right) = -\frac{1}{s^a} Ei\left[\ln(s^a)\right]$$

For a train of spikes alternating at $q = 1, 2, 3, 4 \ldots$, we have
$k(q) = \delta(q-1) - \delta(q-2) + \delta(q-3) - \delta(q-4) + \ldots$
By using the shifted dirac-delta and its transform as $\delta(t - t_0) \leftrightarrow e^{-st_0}$, we get

$$P(r) = e^{-r} - e^{-2r} + e^{-3r} - e^{-4r} + \ldots = (e^{-r} + e^{-3r} + e^{-5r} + \ldots)$$
$$-(e^{-2r} + e^{-4r} + e^{-6r} + \ldots)$$
$$= \frac{e^{-r}}{1 - e^{-2r}} - \frac{e^{-2r}}{1 - e^{-2r}} = \frac{1 - e^{-r}}{e^r - e^{-r}} = \frac{1}{2\sinh(r)} (1 - e^{-r})$$

Substituting $r = -\ln(s)$ we obtain

$$P(s) = \frac{1}{2\sinh[-\ln(s)]} \left(1 - e^{-[-\ln(s)]}\right) = \frac{1}{\sinh[\ln(s)]} \left(\frac{s-1}{2}\right)$$

For $k(q) = \begin{cases} 1 & q \geq 0 \\ 0 & q > 1 \end{cases}$, call this as "window order-one" $WIN(0, 1)$.

Then $P(r) = \int_0^{\infty} 1 \cdot e^{-rq} dq$ is

$$P(r) = \int_0^1 e^{-rq} dq = \frac{1 - e^{-r}}{r}$$

10.5 Analysis of Continuous Order Distribution

and substituting $r = -\ln(s)$ we have

$$P(s) = \frac{1 - e^{-r[-\ln(s)]}}{-\ln(s)} = \frac{s - 1}{\ln(s)}$$

For $k(q) = \begin{cases} 1 & q \geq 0 \\ 0 & q > 2 \end{cases}$, call this as "window order-two" $WIN(0, 2)$, then $P(r) = \frac{1 - e^{-2r}}{r}$, and substituting $r = -\ln(s)$, we have

$$P(s) = \frac{s^2 - 1}{\ln(s)}$$

For order distribution with continuous from 0 to 1 as $WIN(0, 1)$ and spike at $q = 2$ is $k(q) = WIN(0, 1) + \delta(q - 2)$, then
$P(r) = L\{WIN(0, 1)\} + L\{\delta(q - 2)\}$ gives

$$P(r) = \frac{1 - e^{-r}}{r} + e^{-2r} = \frac{1 - e^{-r} + re^{-2r}}{r}$$

and substituting $r = -\ln(s)$ we have

$$P(s) = \frac{1 - e^{-[-\ln(s)]} + [-\ln(s)]e^{-2[-\ln(s)]}}{-\ln(s)} = \frac{s^2 \ln(s) + s - 1}{\ln(s)}$$

For $k(q) = \begin{cases} q & q \geq 0 \\ 0 & q > 2 \end{cases}$, call this as $+RAMP(0, 2)$.

In this derivation, we use Laplace identity: $t^n f(t) \leftrightarrow (-1)^n \frac{d^n F(s)}{ds^n}$.
This observation makes construction of $+RAMP(0, 2)$ from $WIN(0, 2)$ as:

$$+RAMP(0, 2) = qWIN(0, 2)$$

So Laplace will be $qWIN(0, 2) \leftrightarrow (-1)^1 \frac{d}{dr} L\{WIN(0, 2)\}$

$$P(r) = (-1)\frac{d}{dr}\frac{1 - e^{-2r}}{r} = \frac{1 - 2re^{-2r} - e^{-2r}}{r^2} = \frac{1 - (2r + 1)e^{-2r}}{r^2}$$

and substituting $r = -\ln(s)$, we have $P(s) = \frac{1 - (2[-\ln(s)] + 1)e^{-2[-\ln(s)]}}{[\ln(s)]^2} = \frac{1 + [2\ln(s) - 1]s^2}{[\ln(s)]^2}$

For $k(q) = \begin{cases} q & q \geq 0 \\ 0 & q > 1 \end{cases}$, call this as $+RAMP(0, 1)$, using similar procedure as above we get

$$P(r) = \frac{1 - (1 + r)e^{-r}}{r^2}$$

and substituting $r = -\ln(s)$, we get $P(s) = \frac{1+[\ln(s)-1]s}{[\ln(s)]^2}$

For $k(q) = \begin{cases} -q+2 & q \geq 0 \\ 0 & q > 2 \end{cases}$, call this as $-RAMP(0, 2)$, can be composed as:

$$k(q) = -[+RAMP(0, 2)] + 2[WIN(0, 2)],$$

By using derived Laplace of these constituents we get

$$P(r) = -\left(\frac{1 - e^{-2r} - 2re^{-2r}}{r^2}\right) + 2\left(\frac{1 - e^{-2r}}{r}\right) = \frac{2r - 1 + e^{-2r}}{r^2},$$

and substituting $r = -\ln(s)$, we get

$$P(s) = \frac{s^2 - 1 - 2\ln(s)}{[\ln(s)]^2}$$

For $k(q) = \begin{cases} -q+1 & q \geq 0 \\ 0 & q > 1 \end{cases}$, call this as $-RAMP(0, 1)$ and by the above procedure, we compose this and write

$k(q) = -[+RAMP(0, 1)] + 1[WIN(0, 1)]$, then taking Laplace of the constituents we get

$$P(r) = -\left(\frac{1 - (1+r)e^{-r}}{r^2}\right) + 1\left(\frac{1 - e^{-r}}{r}\right) = \frac{r - 1 + e^{-r}}{r^2}$$

and substituting $r = -\ln(s)$, we get $P(s) = \frac{s-1-\ln(s)}{[\ln(s)]^2}$

For $k(q) = \begin{cases} 0 & q \geq 0 \\ -q+2 & q \geq 1 \\ 0 & q > 2 \end{cases}$ is $-RAMP(0, 1)$ shifted from $q = 0$ to $q = 1$.

Here Laplace shift identity $f(t - t_0) \leftrightarrow e^{-st_0} F(s)$ is used to get

$$P(r) = e^{-r}\left(\frac{r - 1 + e^{-r}}{r^2}\right) = \frac{re^{-r} - e^{-r} + e^{-2r}}{r^2} = \frac{(r-1)e^{-r} + e^{-2r}}{r^2}$$

and substituting $r = -\ln(s)$, we get

$$P(s) = \frac{([-\ln(s)] - 1)e^{-[-\ln(s)]} + e^{-2[-\ln(s)]}}{[\ln(s)]^2} = \frac{s^2 + s[\ln(s) + 1]}{[\ln(s)]^2}$$

For $k(q) = \begin{cases} q & q \geq 0 \\ -q+2 & q \geq 1 \\ 0 & q > 2 \end{cases}$ can be composed by

$$+RAMP(0, 1) + [-RAMP(0, 1)]_{at\ q=1}$$

10.5 Analysis of Continuous Order Distribution

From above obtained results for shifted $-RAMP(0, 1)$ and $+RAMP(0, 1)$, we get

$$P(r) = \frac{1 - e^{-r} - re^{-r}}{r^2} + \frac{re^{-r} - e^{-r} - e^{-2r}}{r^2} = \frac{1 - 2e^{-r} + e^{-2r}}{r^2}$$

and substituting $r = -\ln(s)$, we get $P(s) = \frac{1 - 2s + s^2}{[\ln(s)]^2}$

For $k(q) = \frac{1}{2} - \frac{1}{2}\cos(2\pi q)$, we use standard Laplace transform for Heaviside step and cosine expressions to obtain

$$P(r) = \frac{1}{2} \times \frac{1}{r} - \frac{1}{2}\left(\frac{r}{r^2 + 4\pi^2}\right) = \frac{2\pi^2}{r(r^2 + 4\pi^2)}$$

and substituting $r = -\ln(s)$, we get

$$P(s) = \frac{-2\pi^2}{\ln(s)\left\{[\ln(s)]^2 + 4\pi^2\right\}}.$$

We now use this derived expression to truncated $k(q) = \begin{cases} 0.5 - 0.5\cos(2\pi q) & q \geq 0 \\ 0 & q > 2 \end{cases}$; this can be composed by continuous function $\frac{1}{2} - \frac{1}{2}\cos(2\pi q)$ and from this subtract the shifted function at $q = 2$, that is $\frac{1}{2} - \frac{1}{2}\cos[2\pi(q-2)]$. Using shift identity of Laplace operation we get

$$P(r) = \frac{2\pi^2}{r(r^2 + 4\pi^2)} - e^{-2r}\left(\frac{2\pi^2}{r(r^2 + 4\pi^2)}\right) = \frac{2\pi^2(1 - e^{-2r})}{r(r^2 + 4\pi^2)}$$

and substituting $r = -\ln(s)$, we get

$$P(s) = \frac{2\pi^2(s^2 - 1)}{\ln(s)\left\{[\ln(s)]^2 + 4\pi^2\right\}}$$

If $k(q)$ is composed of one continuous function $f(q)$ multiplied by $aWIN(0, q_n)$, then to get r-Laplace transform, convolution identity is used.

Meaning $P(r) = L\{f(q)\} * L\{aWIN(0, q_n)\}$

From the system order distribution, continuous, spiked, or mixed $k(q)$, it is thus possible to obtain the Laplace transform $P(s)$ of the same (by going through intermediate r-Laplace i.e. $P(r)$). The reciprocal of the Laplace transform of the system order distribution gives the frequency response or the system transfer function, $G(s) = 1/P(s)$. When Laplace transform of the input excitation is multiplied by reciprocal of Laplace transform of order distribution, we get the output i.e. $X(s) = F(s)/P(s)$, $F(s)$ is excitation. Conversely by controlling or manipulating the shape/form of $P(s)$, the system's transfer function or the output shape/form can be controlled, with infinite freedom. The manipulation of the order distribution $k(q)$ thus gives a thought for "continuum order feedback controller."

Figure 10.2 represents the system identification method applied for the transfer function $G(s) = \frac{1}{s^2+1.5s^{1.5}+s+1.5s^{0.5}+1}$. Equation (10.20) with $Q = 0.1$, $N = 25$ with ω spaced logarithmically between 10^{-2} and 10^2 gives the discretized plots for $k(q)$, the order distribution. Similarly Fig. 10.3 represents the order distribution obtained for $G(s) = \frac{1}{s^2+0.5s+1}$. The observation in Figs. 10.2 and 10.3 is that the order spikes that one would have obtained are somewhat smeared and are blunt. The techniques to concentrate the $k(q)$ distribution peaks into specific discrete q values is a topic of advance research and development. The concentration of the spikes should ideally—as in Fig. 10.1, for integer as well as fractional order system—be identified. Given order distribution in discrete form, the transfer function is reconstructed by

$$G(s) \equiv \frac{1}{\sum_{n=0}^{N} k_n s^{nQ}}.$$

Sampling issues are described here for different types of order distribution (spiked, continuous, and mixed). In evaluating the order distribution integral, it will be important to distinguish between types, i.e. spiked (impulsive) and continuous. If the composition of the order distribution is assumed to be purely impulsive entirely, or more generally $k(q) = \sum_{n=0}^{N} k_\delta(nQ)\delta(q - nQ)$, $q_{max} = NQ$, then the integral can be evaluated by Euler expansion. The approximation then for the integral is $\int_{0}^{q_{max}} k(q) s^q dq \cong \sum_{n=0}^{N} k_\delta(nQ) s^{nQ} Q$. Here, in this discussion, a note may be taken as it is observed that Q appears explicitly in the summation. The unit impulses are spikes of unit area; their heights when identified by the above method are amplified by $1/Q$. This height gets reduced by Q in the summation to give correct results.

For continuous order distribution (for type say $k(q) = \exp(-q)$) and others in Table 10.1, the integral can be evaluated by inter-sample reconstruction technique. This is analogous to sampled data reconstruction problem in time domain. Assuming

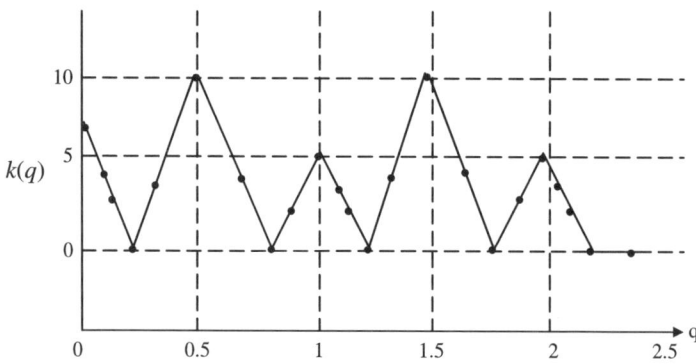

Fig. 10.2 Order distribution for transfer function $G(s) = \frac{1}{s^2+1.5s^{1.5}+s+1.5s^{0.5}+1}$

10.5 Analysis of Continuous Order Distribution

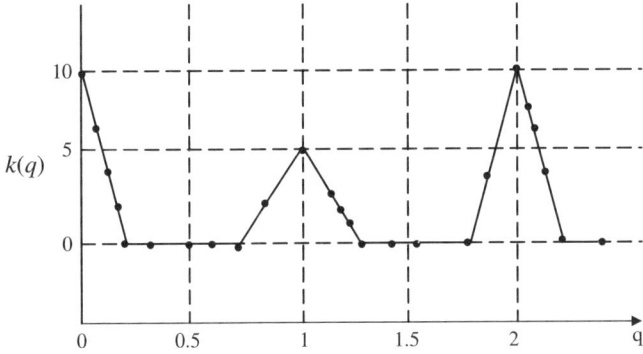

Fig. 10.3 Order distribution of transfer function $G(s) = \frac{1}{s^2+0.5s+1}$

the order distribution to be piecewise constant, then the approximate integral is expressed as $\int_0^{q_{max}} k(q) s^q dq \cong \sum_{n=0}^{N} k_c(nQ) \int_{nQ}^{(n+1)Q} s^q dq = \frac{s^Q-1}{\ln(s)} \sum_{n=0}^{N} k_c(nQ) s^{nQ}$.

Evaluation of $\int_{nQ}^{(n+1)Q} s^q dq$ is obtained by writing $s^q = e^{q\ln(s)}$. So

$$\int_{nQ}^{(n+1)Q} s^q dq = \int_{nQ}^{(n+1)Q} e^{q\ln(s)} dq = \left[\frac{e^{q\ln(s)}}{\ln(s)}\right]_{nQ}^{(n+1)Q} = \left[\frac{s^q}{\ln(s)}\right]_{nQ}^{(n+1)Q}$$

$$= \frac{1}{\ln(s)} \left[s^{(n+1)Q} - s^{nQ}\right] = \frac{s^{nQ}\left[s^Q - 1\right]}{\ln(s)}$$

Notice that Q does not explicitly multiply the summation, though implicitly being present in numerator. Using this discussion, the expression (10.22) is modified and presented as:

$$\begin{bmatrix} \frac{(j\omega_1)^Q-1}{\ln(j\omega_1)} \left\{1 \; (j\omega_1)^Q \; (j\omega_1)^{2Q} * (j\omega_1)^{NQ}\right\} \\ \frac{(j\omega_2)^Q-1}{\ln(j\omega_2)} \left\{1 \; (j\omega_2)^Q \; (j\omega_2)^{2Q} * (j\omega_2)^{NQ}\right\} \\ * \\ \frac{(j\omega_M)^Q-1}{\ln(j\omega_M)} \left\{1 \; (j\omega_M)^Q \; (j\omega_M)^{2Q} * (j\omega_M)^{NQ}\right\} \end{bmatrix} \begin{bmatrix} k_{c0} \\ k_{c1} \\ * \\ k_{cN} \end{bmatrix} = \begin{bmatrix} 1/G(j\omega_1) \\ 1/G(j\omega_2) \\ * \\ 1/G(j\omega_M) \end{bmatrix}$$

If the order distribution is assumed to be with mixed type, that is, with both the continuous order distribution and the impulsive type, then the reconstruction procedure will be mixed for the above discussed procedure, for the system identification experiments. Expression (10.16) gets modified as:

$$\left[\sum_{n=0}^{N} k_\delta(nQ) s^{nQ} Q + \frac{s^Q-1}{\ln(s)} \sum_{n=0}^{N} k_c(nQ) s^{nQ}\right]_{s=j\omega_1} = \frac{1}{G(j\omega_1)}. \quad \text{Expression (10.22)}$$

thus for the system identification with mixed order distribution is

$$\begin{bmatrix} Q\{1\,(j\omega_1)^Q * (j\omega_1)^{NQ}\} & \frac{(j\omega_1)^Q-1}{\ln(j\omega_1)}\{1\,(j\omega_1)^Q * (j\omega_1)^{NQ}\} \\ Q\{1\,(j\omega_2)^Q * (j\omega_2)^{NQ}\} & \frac{(j\omega_2)^Q-1}{\ln(j\omega_2)}\{1\,(j\omega_2)^Q * (j\omega_2)^{NQ}\} \\ * & * \\ Q\{1\,(j\omega_M)^Q * (j\omega_M)^{NQ}\} & \frac{(j\omega_M)^Q-1}{\ln(j\omega_M)}\{1\,(j\omega_M)^Q * (j\omega_M)^{NQ}\} \end{bmatrix} \begin{bmatrix} k_{\delta 0} \\ k_{\delta 1} \\ * \\ k_{\delta N} \\ k_{c 0} \\ k_{c 1} \\ * \\ k_{c N} \end{bmatrix} = \begin{bmatrix} 1/G(j\omega_1) \\ 1/G(j\omega_2) \\ * \\ 1/G(j\omega_M) \end{bmatrix}$$

10.6 Variable Order System

While developing the concept of continuous order distribution, some thought was put in, that is, if the system is subjected to ambient variations, say temperature variation, then the fixed order does show distribution. For instance, taking the transfer impedance of a half-order element at fixed ambience is $Z(j\omega) \propto \frac{1}{(j\omega)^{0.5}}$ may well show a change in order 0.5 with the variable ambience. Therefore, the differintegrals of the fractional or integer order may well have the order q, which varies with time i.e. becomes $q(t)$. Consider the fractional differential equation $_cD_t^q y(t) = f(t)$, and the inferred integral equation is $_cD_t^{-q} f(t) = y(t)$.

Since q can take any real value, the development of calculus for varying q with t and y is essential field or research. The variable order system will have $_cD_t^{-q(t,y)} f(t) = y(t)$.

10.6.1 RL Definition for Variable Order

Consider only time variation of the order, i.e., $q \to q(t)$, the RL definition with zero initial condition $\psi(f, -q(t), a, c, t) = 0$ yields

$$_0D_t^{-q(t)} f(t) \equiv \int_0^t \frac{(t-\tau)^{q_e(t,\tau)-1}}{\Gamma(q_g(t,\tau))} f(\tau) d\tau$$

Here the arguments of exponent are q_e, and the Gamma function argument is q_g may be different. This is the basic difference from fixed-order system. The above formulation of the definition can have $q(t, \tau) \to q(t), q(t, \tau) \to q(\tau), q(t, \tau) \to q(t-\tau)$. Substitution of these three definitions into the above formulation of $q_e(t, \tau)$ and $q_g(t, \tau)$ yields nine expressions of the variable order fractional RL integration. These nine definitions are subjected to the criteria and desirable properties of fractional integration, to rule out the undesirable formulations. The most important ones

10.6 Variable Order System

are linearity and index law (composition), leaving aside backward compatibility and zero property.

Consider the linearity property for the definition of variable order integration as mentioned above:

$$_0D_t^{-q(t)}(af(t)+bg(t)) = \int_0^t \frac{(t-\tau)^{q_e(t,\tau)-1}}{\Gamma(q_g(t,\tau))}\{af(\tau)+bg(\tau)\}d\tau$$

$$= a\int_0^t \frac{(t-\tau)^{q_e(t,\tau)-1}}{\Gamma(q_g(t,\tau))}f(\tau)d\tau$$

$$+ b\int_0^t \frac{(t-\tau)^{q_e(t,\tau)-1}}{\Gamma(q_g(t,\tau))}g(\tau)d\tau$$

$$= a\,_0D_t^{-q(t)}f(t) + b\,_0D_t^{-q(t)}g(t)$$

The above derivation shows that linearity is satisfied for all arguments of q in numerator as well as denominator for the definition of fractional integral.

Out of all these substitutions, the definition of the variable order fractional RL integration when $q_e(t,\tau) = q_g(t,\tau)$ gives interesting observation, for the substitution case with, $q(t,\tau) \to q(t-\tau)$, which provided adherence to the index law, but failed to satisfy composition law under the following definition:

$$_0D_t^{-q(t)}f(t) \equiv \int_0^t \frac{(t-\tau)^{q(t-\tau)-1}}{\Gamma(q(t-\tau))}f(\tau)d\tau$$

Here, it was inferred that $_0D_t^{-q(t)}\,_0D_t^{-v(t)}f(t) = \,_0D_t^{-v(t)}\,_0D_t^{-q(t)}f(t) \neq \,_0D_t^{-q(t)-v(t)}f(t)$. The detailed proof uses the convolution theory and the convolution nature of this particular definition, which appears to be the most satisfactory one for use.

It should be mentioned that different physical processes might effectively use different definitions, and all the three forms, namely,

$$_0D_t^{-q(t)}f(t) \equiv \int_0^t \frac{(t-\tau)^{q(t-\tau)-1}}{\Gamma(q(t-\tau))}f(\tau)d\tau,$$

$_0D_t^{-q(t)}f(t) \equiv \int_0^t \frac{(t-\tau)^{q(t)-1}}{\Gamma(q(t))}f(\tau)d\tau$, and $_0D_t^{-q(t)}f(t) \equiv \int_0^t \frac{(t-\tau)^{q(\tau)-1}}{\Gamma(q(\tau))}f(\tau)d\tau$ may prove useful.

This is because this formulation adheres to the index law and because of the convolution forms, which makes available all of the results of the associated theory, the most compelling definition is

$$_0D_t^{-q(t)}f(t) \equiv \int_0^t \frac{(t-\tau)^{q(t-\tau)-1}}{\Gamma(q(t-\tau))} f(\tau)d\tau \quad q(t) > 0$$

Figure 10.4 describes a variable order structure.

If the above form of equations are assumed, then parallel definition for the variable structure derivative might be considered as:

$$_0D_t^{q(t)}f(t) \equiv {_0D_t^m} {_0D_t^{-p(t)}}f(t), \quad q(t) > 0$$

where $q(t) = m - p(t)$, and m is positive integer. If it is assumed that $q(t)$ be always positive, then m could be taken as the least integer greater than $q(t)$. However, since composition does not hold, it is not clear that if this could be a reasonable definition. Matters are further complicated by the fact that it may be desirable to allow $q(t)$ to range over both positive and negative values. This places a "seam" at $q = 0$, which may make any approach based on RL definition implausible, and perhaps requires an approach based on GL definition.

10.6.2 Laplace Transforms and Transfer Function of Variable Order System

The derivation of the Laplace transforms of the variable order structure integral follows that for the fixed-order case exactly, since the convolution theorem can be applied. Then, considering the un-initialized case of fractional integration of variable structure, we get

$$L\left\{_0D_t^{-q(t)}f(t)\right\} = \int_0^\infty e^{-st}\left(\int_0^t \frac{(t-\tau)^{q(t-\tau)-1}}{\Gamma(q(t-\tau))}f(\tau)d\tau\right)dt, \quad q(t) > 0, t > 0$$

With definition of convolution $L\{h(t)*g(t)\} = H(s)G(s) = L\left(\int_0^t h(\tau)g(t-\tau)d\tau\right)$, and taking $h(t) = f(t)$ and $g(t) = \frac{t^{q(t)-1}}{\Gamma(q(t))}$, the convolution theorem yields

$$L\left\{_0D_t^{-q(t)}f(t)\right\} = F(s)G(s) = L\{f(t)\}L\left\{\frac{t^{q(t)-1}}{\Gamma(q(t))}\right\}$$

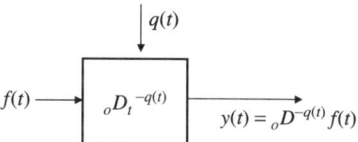

Fig. 10.4 Variable order structure

10.6 Variable Order System

The variable order structure differintegration allows the introduction of a new transfer function concept. Refer Fig. 10.4; the conventional transfer function relates the Laplace transform of the output to the transform of the input (ratio of the two) by:

$$TF_1 \equiv \frac{L\{y(t)\}}{L\{f(t)\}} = \frac{L\left\{{}_0D_t^{-q(t)}f(t)\right\}}{L\{f(t)\}} = \frac{L\{f(t)\}L\left\{\frac{t^{q(t)-1}}{\Gamma(q(t))}\right\}}{L\{f(t)\}} = L\left\{\frac{t^{q(t)-1}}{\Gamma(q(t))}\right\}$$

Since the $q(t)$ is now a variable, it may be thought of as an input (Fig. 10.4), and for this, a new transfer function may be defined as $TF_2 \equiv \frac{L\{y(t)\}}{L\{q(t)\}}$. For the considered definition, the process from $q(t)$ to $y(t)$ has not been shown to be linear.

That is, ${}_0D_t^{-q_1(t)-q_2(t)}f(t) = {}_0D_t^{-q_1(t)}f(t) + {}_0D_t^{-q_2(t)}f(t)$ has not been shown. Here further research is required to define and utilize new transfer function (TF_2).

The relationship of the two transfer functions may be determined as follows. Consider $f(t)$ and $q(t)$ to be related by $g(t)$, where
$q(t) = \int\limits_0^t f(t)g(t-\tau)d\tau$, then $L\{q(t)\} = L\{f(t)\}L\{g(t)\}$, by convolution theorem. Then:
$TF_2 = \frac{L\{y(t)\}}{L\{q(t)\}} = \frac{L\{f(t)\}TF_1}{L\{f(t)\}L\{g(t)\}}$ gives $\frac{TF_1}{TF_2} = L\{g(t)\}$

10.6.3 GL Definition for Variable Order

A variable structure differintegral is formed by GL definition too. Let

$$\Delta T = (t-a)/N$$

and limit consideration to $q \to q(t)$ or $q \to q(t - j\Delta T)$. Then expressing q generally as $q \to q(t, j\Delta T)$, a generalized GL form may be written as:

$${}_aD_t^{q(t)}f(t) = \lim_{\substack{N \to \infty \\ \Delta T \to 0}} \sum_{j=0}^{N-1} \Delta T^{q_E(t, j\Delta T)} \frac{\Gamma(j - q_N(t, j\Delta T))}{\Gamma(-q_D(t, j\Delta T))\Gamma(j+1)} f(t - j\Delta T)$$

It is observed that $q(t, j\Delta T)$ occurs three times in this expression, in the exponent as q_E, in the numerator as q_N, and in the denominator as q_D. Thus, the number of formula combination (permutation) is eight, as compared to nine in RL type definition previously discussed.

For the combinations $q_E(t, j\Delta T) \to q_E(t - j\Delta T)$, $q_N(t, j\Delta T) \to q_N(t - j\Delta T)$, $q_D(t, j\Delta T) \to q_D(t - j\Delta T)$, the GL formulation is as follows:

$${}_aD_t^{q(t)}f(t) = \lim_{\substack{N \to \infty \\ \Delta T \to 0}} \sum_{j=0}^{N-1} \Delta T^{q_E(t-j\Delta T)} \frac{\Gamma(j - q_N(t - j\Delta T))}{\Gamma(-q_D(t - j\Delta T))\Gamma(j+1)} f(t - j\Delta T)$$

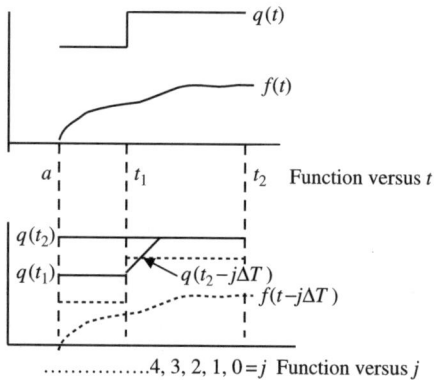

Fig. 10.5 Variable $q(t)$ differintegration by GL method

This is the most promising definition out of the eight permutations.

Consider two time interval evaluations of GL differeintegration of $_aD_t^{q(t)}f(t)$, for common $q(t)$ with one evaluation for $a < t \leq t_1$, and another in interval $a < t \leq t_2$, where $t_2 > t_1$ (Fig. 10.5). The function did not exist before the time $t = a$, so $f(t) = 0$ for $t < a$. In particular, consider the order $q(t)$ steps from constant value $q(t_1) \to q(t_2)$, a new constant value at time $t = t_1$. The evaluation of $_aD_t^{q(t)}f(t)$ to $t = t_2$ can be viewed as an evaluation from $t = a \to t_1$, summed to an evaluation from $t = t_1 \to t_2$. In view of the hereditary nature of the fractional differintegration, the evaluation to $t = t_1$ is part of history in the evaluation a to t_2; thus it is apparent that for the evaluation of $_aD_t^{q(t)}f(t)$ to $t = t_2$ based on the value $q_E(t, j\Delta T) \to q_E(t - j\Delta T)$, $q_N(t, j\Delta T) \to q_N(t - j\Delta T)$, and $q_D(t, j\Delta T) \to q_D(t - j\Delta T)$ will yield desirable results; all others will give undesirable result. This is also because, over the period $t < t_1$, part of evaluation, the value of $q(t, j\Delta T) = q(t) = q(t_2)$ will be used in the summation, essentially changing history. Thus, it appears that the only satisfactory definition is as mentioned above. This definition parallels the RL promising formulation, and adds credibility of convolution-related variable structure form. The intuitive nature of GL form, in terms of available conceptualization, allows the hope that a form of it may evolve, which would satisfy the composition property. The detailed study in this direction is still a matter of research.

10.7 Generalized PID Controls

Here in this section generalization of the PI-PID controller is presented. The generalization is possible only because of the availability of the fractional order elements. The standard integer order PI-controller has a transfer function of the form:

$$H(s) = k_p + \frac{k_i}{s} \tag{10.24}$$

10.7 Generalized PID Controls

This can be written as:

$$H(s) = k_p + k_i s^{-1} \qquad (10.25)$$

The controller can now be generalized using fractional order integrals. Assuming the base fraction $q = 1/N$, for now, a generalized PI-controller can be written as:

$$H(s) = k_0 + k_1 s^{-q} + k_2 s^{-2q} + \ldots + k_{N-1} s^{-(N-1)q} + k_N s^{-1}, \quad 0 < q < 1 \qquad (10.26)$$

$$H(s) = \sum_{n=0}^{N} k_n s^{-nq}, \, Nq = 1 \qquad (10.27)$$

Performing manipulation on above expressions we obtain

$$H(s) = \frac{k_0 s + k_1 s^{(N-1)q} + \ldots + k_{N-1} s^q + k_N}{s}, \, Nq = 1 \qquad (10.28)$$

$$H(s) = \frac{\sum_{n=0}^{N} k_n s^{(N-n)q}}{s}, \quad Nq = 1 \qquad (10.29)$$

Clearly, this controller allows a much degree and much variety of compensation results. Inserting this into the standard closed-loop control configuration with plant transfer function $G(s)$ gives closed-loop transfer function

$$T(s) = G(s)H(s)/(1 + G(s)H(s))$$

as:

$$T(s) = \frac{\left[\sum_{n=0}^{N} k_n s^{(N-n)q}\right] G(s)}{s + \left[\sum_{n=0}^{N} k_n s^{(N-n)q}\right] G(s)}, \quad Nq = 1 \qquad (10.30)$$

With $G(s) = N(s)/D(s)$, the close-loop transfer function becomes

$$T(s) = \frac{\left[\sum_{n=0}^{N} k_n s^{(N-n)q}\right] N(s)}{s D(s) + \left[\sum_{n=0}^{N} k_n s^{(N-n)q}\right] N(s)} \qquad (10.31)$$

This generalization of a PI-control process gives considerable design capability and freedom. Both close-loop poles and close-loop zeros can be placed by proper selection of the gains. The compensator can be further generalized by considering the powers of q to be unrelated; in this case, the controller is

$$H(s) = k_0 + k_1 s^{-q_1} + k_2 s^{-q_2} + \ldots + k_{N-1} s^{-q_{N-1}} + k_N s^{-1} = \frac{\sum_{n=0}^{N} k_n s^{q_{N-n}}}{s} \quad (10.32)$$

The summation assumes $q_0 = 0$ and $q_N = 1$, and $0 \leq q_i \leq 1$ and $0 \leq i \leq N$.

The above-discussed approach can be extended to PID controllers, where derivative action controller is allowed. However, the derivative control is seldom used because its high-frequency attenuation of noise issues. Still to generalize the concept, the PID controller can thus be generalized as:

$$H(s) = k'_N s^{+1} + k'_{N-1} s^{+(N-1)q} + \ldots + k'_2 s^{+2q} + k'_1 s^{+q} + k_0 + k_1 s^{-q}$$
$$+ k_2 s^{-2q} + \ldots + k_{N-1} s^{-(N-1)q} + k_N s^{-1} \quad (10.33)$$

or equivalently with $q = 1/N$, the generalized PID is $H(s) = \sum_{n=-N}^{N} h_n s^{nq}$. Based on the previous discussions, the same compensator can be decomposed into numerator and denominator and be expressed as $H(s) = \frac{\sum_{n=0}^{N} b_n s^{n/N}}{\sum_{n=0}^{N} a_n s^{n/N}} = \frac{\sum_{n=0}^{N} b_n s^{P_n}}{\sum_{n=0}^{N} a_n s^{q_n}}$. In a closed-loop feedback configuration, the plant $G(s) = N(s)/D(s)$, we get the generalized transfer function:

$$T(s) = \frac{\left[\sum_{n=0}^{N} b_n s^{P_n}\right] N(s)}{\left[\sum_{n=0}^{N} a_n s^{q_n}\right] D(s) + \left[\sum_{n=0}^{N} b_n s^{P_n}\right] N(s)} \quad (10.34)$$

To determine the effectiveness of the generalized PID controller, the analysis must be done on Nyquist plane, the quality of which depends on the approximation limited by approximation size.

10.8 Continuum Order Feedback Control System

As discussed in system identification section, taking the summation to the limit, we get the transfer function of (10.33) in the form:

$$H(s) = \frac{\int_a^b K_N(q) s^q dq}{\int_a^b K_D(q) s^q dq} \quad (10.35)$$

This is continuum order compensator, where the order distribution $K(q)$ must be selected for numerator and denominator so that $H(s)$ remains causal and so that the

10.8 Continuum Order Feedback Control System

integrand must converge. The general PID expression (10.33) can be expressed as power series as

$$H(s) = \frac{\sum_{n=0}^{N} b_n s^{p_n}}{\sum_{n=0}^{N} a_n s^{q_n}} = \sum_{n=0}^{\infty} c_n s^{r_n}$$

or asymptotic series as $= \sum_{n=0}^{\infty} c_n s^{-w_n}$ representations. Thus, the power integral representation for the continuum compensator would look as:

$H_{PS}(s) = \int_0^a K_{PS}(q) s^q dq$. And the asymptotic series would be

$$H_{AS}(s) = \int_0^\infty K_{AS}(q) s^q dq.$$

The combined form will be having a form as $H(s) = \int_{-\infty}^{a} K(q) s^q dq$, where the order distribution function $K(q)$ is chosen to make the integral convergent. For an order distribution for generalizing the PID controller of (10.33) will be

$$H(s) = \int_{-1}^{+1} K(q) s^q dq \tag{10.36}$$

Given the $K(q)$ function, it is easy to perform the analysis of closed-loop system in Nyquist plane. The selection of suitable $K(q)$ is field of research; following example of an oscillating plant will give some insight of the open issues.

Consider a plant transfer function $G(s) = \frac{1}{s^2+1}$ (an oscillator un-damped). A possible compensator using the order distribution would be $H(s) = \int_0^2 K(q) s^q dq$, and if this controller were placed before the plant, the resulting close-loop system will have a transfer function, as

$$T(s) = \frac{\int_0^2 K(q) s^q dq}{s^2 + \int_0^2 K(q) s^q dq + 1};$$

this compensator allows an infinite number of frequencies in the closed-loop system and thus allows considerable freedom to design the appropriate $K(q)$. The question of selection of this function is to minimize some desired error, for a given input, or more appropriately a cost function, but much research is required in this direction.

10.9 Time Domain Response of Sinusoidal Inputs for Fractional Order Operator

Replacing $s^q \to (j\omega)^q$ gives frequency response of the transfer function. Frequency domain approaches assume that the time responses are sinusoidal steady state. Whenever, an input is applied to a system, the response will consists of transient part and steady-state part. The frequency response approach assumes that the transient has decayed away, and that the response is in sinusoidal steady state. For fractional order systems, sinusoid steady state also implies that the initialized response due to initialization function (ψ) has decayed to near zero.

A periodic function with period T is represented as Fourier series:
$f(t) = \sum_{k=1}^{\infty} \left(c_k e^{j2\pi kt/T} + \bar{c}_k e^{-j2\pi kt/T} \right)$. Fourier integral obtains the coefficient, and the coefficients are complex conjugate. The coefficient is

$$c_k = \frac{1}{T} \int_0^T f(t) e^{-j2\pi kt/T} dt.$$

Oldham Spanier gives the method for evaluating fractional differentigration of repeated periodic function by using lower- incomplete Gamma function as $_0 d_t^q \left(\exp(\pm j2\pi k/T) \right) = \left(\frac{\pm j2\pi k}{T} \right)^q \gamma(-q, \pm j2\pi kt/T)$. This term contains both the transient and the steady-state fractional derivative of the periodic function. An asymptotic expansion for larger values of t for incomplete Gamma function, γ term, gives

$$_0 d_t^q f(t) = \sum_{k=1}^{\infty} \left(\frac{2\pi k}{T} \right)^q (c_k \exp[j2\pi \{(kt/T) + (q/4)\}] + \bar{c}_k \exp[-j2\pi \{(kt/T) + (q/4)\}])$$

Defining the radian frequency $\omega_0 = 2\pi/T$ gives an equivalent response as:

$$_0 d_t^q f(t) = \sum_{k=1}^{\infty} (k\omega_0)^q \left(c_k \exp j [k\omega_0 t + (\pi q/2)] + \bar{c}_k \exp\{-j [k\omega_0 t + (\pi q/2)]\} \right)$$

For any given input frequency $k\omega_0$, it can be seen that the magnitude of the corresponding fractionally differintegrated steady-state output sinusoid has its magnitude scaled by $(k\omega_0)^q$, and is phase shifted by $q\pi/2$, for example, after the decay of the transient $_0 d_t^q (\sin \omega t) \to \lim_{t \to \infty} (\omega)^q \sin(\omega t + [\pi q/2])$.

This result generalizes the response obtained for integer order systems. Here it is important to note that frequency response results require sinusoidal steady state, and initialization function plays less important role.

10.10 Frequency Domain Response of Sinusoidal Inputs for Fractional Order Operator

In earlier section, it was mentioned that replacement of $s^q \to (j\omega)^q$ gives the steady-state response of the transfer function. However, substitution $s^{nq} \to (j\omega)^{nq}$ for q as a fractional number and nq not necessary an integer, points towards multiple solutions and several roots. The question here is actually which root of $(j)^{nq}$ to use. The primary root is considered to be one with the smallest angle from the positive real axis, with remaining roots being secondary roots. The answer to this question is given by time domain result of the previous section. That is, using the primary roots given by the frequency domain substitution, $s^{nq} \to (j\omega)^{nq}$ will give the frequency response corresponding to the correct time domain response. For example, $s^{0.5}$ substituted by $s = j\omega$ gives which is $j^{0.5}\omega^{0.5}$. Recognizing $\sqrt{j}$ has roots $\exp(j\pi/4)$ and $\exp(j5\pi/4)$, the primary root is always chosen for the frequency response i.e. the root $e^{j\pi/4}$. This observation allows using the standard tools for control system analysis as Bode plot, Nyquist plot, and others.

The frequency response of fractional order differintegral gives insight into the use of the control system tools. The un-initialized Laplace relation is

$$L\left\{{}_0d_t^q f(t)\right\} = s^q F(s).$$

Thus the transfer function of the fractional operator is $H(s) = \frac{L\{{}_0d_t^q f(t)\}}{F(s)} = s^q$. To obtain frequency response replace $s^q \to (j\omega)^q$.

The magnitude response is simply $|H(j\omega)| = \omega^q$, which rolls off at $20q$ dB/decade on Bode plot, and the phase shift is given by the angle of the primary root of $(j)^q$ which is $\arg H(j\omega) = q\pi/2$, the derivative operation for $q > 0$ and integration operation for $q < 0$.

10.11 Ultra-damped System Response

The properties of temporal behavior of systems were discussed in Chap. 7, with respect to pole-locations in w-plane. In Chap. 7 by lines with angle $q\pi$ in w-plane, are on the secondary Riemann sheet of the s-plane. These pole properties were termed as hyper-damped, as they were damped more than usual integer-order over-damped poles. Now with respect to the w-plane, it is with some necessity that distinction is made, between the poles that are on the negative real axis of w-plane (which are called ultra-damped) and complex conjugate poles of w-plane (which are still hyper-damped).

An ultra-damped system will consist of parallel combination of the form

$$H(s) = \frac{Y(s)}{U(s)} = \frac{k}{s^q + a}, a > 0$$

and real. Here k is system gain, a is ultra-damped pole in w-plane and $q > 0$. For simplicity, take $a = 1$, $k = 1$, and frequency response transfer function is $H(j\omega) = \frac{1}{(j\omega)^q + 1}$. For small values of frequency $H(j\omega) = 1$, and magnitude is thus unity and phase angle is zero. For large values of frequency, the transfer function becomes $H(j\omega) = (j\omega)^{-q}$, and frequency response reverts back to that of a simple fractional order operator discussed in the Sect. 10.10.

10.12 Hyper-damped System Response

Hyper-damped systems have a pair of complex conjugate poles off the negative real axis of w-plane, but are farther to the left than the under-damped region (Chap. 7). These poles are at $|\phi| > \pm q\pi$ lines. Here there are several types of behaviors depending upon the specific location of the poles in w-plane.

First, we address a system that can be realized with passive energy storage element. A passive energy storage element is one that cannot return more energy to the system than when placed into the element by the system in the past. An active element is one that can return more energy to a system than when placed into it in past. Typically, an active element will have associated with it, either a large gain or negative gain. Necessary (but not sufficient) condition for fractional order system to be passive are that minimal transfer function denominator have all positive coefficients, and that all poles lie to the left of stability wedge in w-plane. Another concept traditionally associated with passivity is positive real concept. To be positively real system, the frequency response of transfer function must always lie in the right half of Nyquist plot, meaning that the $\arg H(j\omega) \leq \pm \pi/2$ and $\Re e H(j\omega) > 0$. This means that maximum phase shift of positive real transfer function is bounded by $\pm 90°$. A minimum phase system has smallest possible phase shift for a given magnitude response. An implication of this in integer order system is that all the system poles and zeros must lie in the left-half of the s-plane. For fractional order system, the implication of this being minimum phase all pole-zero lie left of instability wedge of w-plane, $|\phi| > \pm q\pi/2$. Consider as an example all pole system, passive and positive real, minimum phase, transfer function $H(s) = \frac{1}{s + as^{0.5} + 1}$. (A fractional order system can be of minimum phase without being positive real.) The properties of this transfer function depend on the pole location as decided by the value of a (Table 10.2).

The under-damped active region for $0 > a > -\sqrt{2}$ has further consideration. The negative value of "a" is indicative of an active system, while w-plane poles remain in "under-damped" region. The fact that there exist under-damped poles implies that this system has a resonance, and a resonance peak should appear in the frequency response Bode diagram too. Thus even though the high-frequency asymptotes, as well as Laplace transformed transfer function, of this example, indicate that this system is only a first order, it can still go into resonance. Clearly, adding more fractional order terms with smaller value of q would allow even more resonance.

10.13 Disadvantage of Fractional Order System

Table 10.2 Pole placements (in w and s -planes) for a first order transfer function with one 'half' order fractional term

Value of a	$w = \rho \exp j\phi$ w-plane	$s = r \exp j\theta$ s-plane	Property
$a < -\sqrt{2}$	$\Re e(w) > 0$, $|\phi| < \frac{\pi}{4}$	$\Re e(s) < 0$, $|\theta| < \frac{\pi}{2}$ Right Half Plane	UNSTABLE
$0 > a > -\sqrt{2}$	$\Re e(w) > 0$, $\frac{\pi}{4} < |\phi| < \frac{\pi}{2}$	$\Re e(s) < 0$, $\frac{\pi}{2} < |\theta| < \pi$ Left Half Plane	STABLE UNDERDAMPED MIN-PHASE
$2 > a > 0$	$\Re e(w) < 0$, $\frac{\pi}{2} < |\phi| < \pi$	$\pi < |\theta| < 2\pi$, Secondary Riemann Sheet	STABLE HYPERDAMPED
$a > 2$	$|\phi| > \pi$	$|\theta| > 2\pi$, Secondary Riemann Sheet	STABLE ULTRADAMPED

Consequently, it appears that the highest power of Laplace variable in transfer function is no longer an indicator of the order effective order of fractional order system, or of number of resonance to expect in its frequency response. We call the resonance as fractional resonance.

10.13 Disadvantage of Fractional Order System

This section points out to some of the disadvantages that specially occur in computational efforts, to realize fractional order systems. This area of science is an evolving field; presently, disadvantages are pointed in order to make distinction between classical integer order calculus. The points are summarized below:

a. Fractional order differential equations accumulate the entire information of the function in weighted form.
b. Fractionally differentiated state variables must be known as long as the system has been operated. This is known as initialization function.
c. For integer order systems, the initialization function is constant and for fractional order systems it is time varying.
d. Integer order system set of state along with system equations is sufficient to predict the response.
e. The fractional dynamic variables do not represent the state of the system.
f. Fractional dynamics require history of states or sufficient number of points by short-memory principle, for initialization function computation.
g. The above memory effect requires large memory. The evolving developments to reduce this requirement in form of power series expansion and continued fraction expansion of the generating function in digital domain is an ongoing process.

10.14 Concluding Comments

A very advanced topic is touched upon as on today, about the reality of an order of a system. Can there be total certainty about the fixed integer order definition of a system, or should the order of a system be spread around principal orders along with several fractional orders, to take care of distributed effects parametric spreads and other realities, are the questions for scientists and engineers of today to think and answer. When order of a system has distribution, then having controller with distributed order may be wise to have for efficient governance. Also ambience can change the order of the system, and future development in variable fractional order system mathematics is an enriched area of research.

Bibliography

1. A.D.Poularikas, *The handbook of formulas and tables for signal processing*, The Electrical Engineering Handbook Series, CRC Press and IEEE Press, New York 1999.
2. A. Erdelyi (ed), *Tables of Integral Transforms*, vol. 1, McGraw-Hill, 1954.
3. M.Gopal, *Modern Control System Theory*, Second Edition April, 1993, Wiley Eastern Limited.
4. F. F. Kuo, *Network Analysis and Synthesis*, Willey 1966.
5. Igor Podlubny, *Fractional differential equations*, Academic Press San Diego 1999.
6. B.J.Lurie, *Three parameter tunable Tilt-Integral-Derivative TID Controller*, US Patent, 5-371-670, 1994.
7. J.C. Wang, *Realization of generalized Wartburg impedance with RC ladder networks and transmission lines*, J. of Electrochem. Soc., vol. 134, no. 8 August 1987, pp. 1915–1920.
8. A. Tustin, et al. *The design of systems for automatic control of the position of massive objects*, The proceedings of Institute of Electrical Engineers, 105C(1), 1958.
9. I.Podlubny *Fractional order systems and $PI^\lambda D^\mu$ controllers,* IEEE Trans. Auto Cont., vol. 44 no. 1, pp 208–214 Jan 1999.
10. I.Podlubny, *Matrix approach to discrete fractional calculus*, Fractional Calculus and Applied Analysis vol. 3, no. 4 (2000).
11. I Podlubny, *The Laplace Transform Method for Linear Differential Equations of the Fractional Order*, Inst. Exp. Phys. Slovak Acad. Sci., no. UEF-02-94, 1994, Kocice.
12. Y.Y.Tsao, B.Onaral and H H Sun, *An algorithm for determining global parameters of minimum phase systems with fractional power spectra.* IEEE Trans. Inst. And Meas (1989), vol. 38, no. 3, pp 723–729.
13. R.L Bagley *The thermorheological complex materials*, Int J. Engg. Sci, vol. 29, no 7, pp 797–806 (1991).
14. A. Carpinteri and F. Mainardi, *Fractals and Fractional Calculus in Continuum Mechanics*, Springer Verlag, Vienna-New York (1997).
15. R.L Bagley and R A Calico, *Fractional order state equations for control of viscoelastic structures*, J. Guid. Cont. and Dyn, vol. 14, no 2, pp 304–311, March-April (1991).
16. R L Bagley and P J Torvik, *Fractional calculus in transient analysis of viscoelastically damped structures*, AIAA Journal, vol. 23, no. 6. pp. 918–925, (1985).
17. C.A. Halijak, *An RC impedance approximation to $(1/s)^{1/2}$*. IEEE Circuit Theory, vol. CT-11, no 4, pp 494–495 Dec 1964.
18. Carlson and Halijak, *Simulation of the fractional derivative operator $\sqrt{s}$ and fractional integral operator $1/\sqrt{s}$*. Kansas State University Bulletin, vol. 45, pp 1–22 July 1961.
19. Carlson et al. *Approximation of fractional capacitors $(1/s)^{1/n}$ by regular Newton process,* IEEE Trans. on Circuit Theory, vol. 11, no. 2, 1964, pp. 210–213.
20. M. Sugi, Y. Hirano, Y.F. Miura, K. Saito, *Simulation of fractal immitance by analog circuits: an approach to optimized circuits*, IEICE Trans. Fundamentals, vol. E82-A, no. 8, 1999, pp. 1627–1635.

21. M. Sugi et al. $\sqrt{\omega}$ -*Variations of AC admittance in the inhomogeneous distributed RC lines.* Jpn. J. Appl. Phys. Vol. 39, no 9A, 2000, pp. 5367–5368.
22. J.Dostal: *Operational Amplifiers*, Butterworth-Heinemann, Boston, 1993.
23. S.C. Dutta Roy, *On realization of constant-argument immitances of fractional operator*, IEEE Trans. on Circuit Theory, vol. 14, no. 3, 1967, pp. 264–374.
24. S. Manabe, *The Non-Integer Integral and its application to Control Systems*, ETJ of Japan, vol. 6, no. 3–4, 1961, pp. 83–87.
25. M. Nakagava, K. Sorimachi, *Basic characteristics of fractance device*, IEICE Trans. fundamentals, vol. E75-A, no. 12, 1992, pp. 1814–1818.
26. H W Bode, *Network analysis and feed back amplifier design*, Van Nostrand New-York 1949.
27. K.B.Oldham and J.Spanier, *The replacement of Fick's law by formulation involving semi differentiation*. J.Electroanal.Chem, vol. 26, pp.331–341 (1970).
28. D. Matignon. *Generalized Fractional Differential and Difference Equation: Stability Properties and Modeling Issues*, Proc. of Maths. Theory of Networks and Systems Symposium, Padova, Italy 1998.
29. K.B.Oldham *Interrelation of current and concentration at electrodes*, J.Appl Electrochem, vol. 21, pp. 1068–1072 (1991).
30. A.M.A El Sayeed *Fractional order diffusion wave equation* Int. J. of Theor. Phy 1996, vol. 35, pp. 311–322.
31. R L Bagley, *Power law and fractional calculus model of viscoelasticity,* AIAA Journal, vol. 27, no. 10, 1989, pp. 1412–1417.
32. I. Podlubny, *Fractional derivatives: A new stage in process modeling and control,* 4th International DAAAM Symposium, Brno, Czech Republic, 16–18 Sept, 1993, pp. 263-264.
33. S.Saha Ray and R K Bera, *Analytical solution of dynamic system containing fractional derivative of order one-half by Admonian decomposition method.* ASME J. of Applied Mechanic July 2005, vol. 72/1.
34. N.Engheta, *The fractional curl operator in electromagnetic.* Microwave Opt.Tech.Letter vol. 17, pp 86–91, 1998.
35. Naqvi Q A and M Abbas, *Fractional duality solutions and corresponding sources.* PIER vol. 25, pp 290–294 (2000).
36. Benson, *The fractional order governing levy motion.* Water resources research, vol. 36 no. 6 pp 1413–1423 June2000.
37. Heymans and Podlubny et al, *Physical interpretation of initial conditions for fractional differential equation with Reimann-Liouvelli fractional derivative.* Rheological Acta (online) Nov29, 2005.
38. Yoshinoi Akoi, Mihir Sen et al *Approximation of transient temperature in complex geometries using fractional derivatives.* Department of Aerospace and Mechanical Engineering University of Notre Damme. Technical Note Dec 19, 2005.
39. Adam Loverro, *Fractional Calculus history definition and application.* Department of Aerospace and Mechanical Engg. University of Notre Damme. Technical Note May 8–2004.
40. Dorcak, Ivo Petras et al, *Fractional order state space models.* Int. Carpathian Control Conference ICCC Malenovice Czech Rep. May 27-30 2002 pp 193–198.
41. Ivo Petras *The fractional order controller method for their synthesis and application.* BERG University Koissce Slovak Republic Technical Note 2004.
42. Ivo Petras, Vinagre, *Practical Applications of digital fractional order controller to Temperature Control*, Acta Montanistica, Slovaca Rocnik 7(2002), 2 pp 131–137.
43. Valimaki et al, *Principles of fractional delay filters* IEEE Conf on Acoustic Speech Processing (ICASSP) Istanbul, Turkey 5-9 June 2000.
44. P. Launsse, A. Oustaloup et al *A restricted complexity controller with CRONE control system design and tuning.* University Bordeaux-ENSEIRB 2003 Technical Note.
45. A. Oustaloup et al. *Frequency band complex non-integer differentiator characterization and synthesis.* IEEE Trans. on circuits and systems-I: Fundamental Theory and Application, vol. 47, no. 1, pp.-25–39, 2000.

46. A. Oustaloup, *From fractality to non-integer order derivation through recursivity, a property common to these two concepts: a fundamental idea for a new process control strategy*, Proceedings of the 12th IMACS World Congress Paris July 18–22, 1988, vol. 3, pp. 203–208.
47. I. Podlubny et al, *Using continued fraction expansion to discretize fractional order derivative*, FDTA Non Linear Dynamic 2003. For submission to special issue on Fractional Derivative and their application.
48. I. Podlubny, Dorcak, Kostial. J, *On fractional derivatives, fractional order dynamic systems and $PI^\lambda D^\mu$ -controllers*, Procc. of the 36th IEEE CDC San Diego December 2004, pp 4985–4990.
49. Chien Chang Tseng (Senior member IEEE), *Improved design of digital fractional order differentiator using fractional sample delay* TCAS 1-1815, 2004.
50. Ivo Petras, *Control of fractional order Chua's system* Univ of Koisscs Slovak Technical Note 2000.
51. Bohennan Gary, *Analog Realization of fractional order control element revisited* Department of Physics Montonna State Univ. Technical Note Oct 27, 2002.
52. Yang Quan Chen *Analytical stability bound for class of delayed fractional order dynamic system*, IEEE Conf. On Decision and control Florida WeA05-1 2001 pp 1421–1424.
53. Kundert.K, *Modeling di-electric absorption in capacitors* The Designer Guide Community Technical Note 2004.
54. Tom Hartley, Carl F Lorenzo, H. K. Qammer, *Chaos in fractional Chua's circuit*, IEEE Trans. On Circuits and Systems-I: Fundamental Theory and Applications, vol. 42, no. 8, 1995, pp. 485–490.
55. Kiran M Kolwankar, Anil D Gangal, *Local Fractional Derivatives and fractal functions of several variables*, J of Physics 9801010 vol. 1 1998.
56. Kolwankar and Gangal, *Fractional differentiability of nowhere differentiable functions and dimensions*, Chaos 6(4) pp. 505–513, (1996).
57. K S Miller and Ross, *An Introduction to Fractional Calculus and Fractional Differential Equations*, Willey New York 1993.
58. K.B.Oldham and J.Spanier, *Fractional calculus* Academic Press San Diego 1974.
59. H M Srivastava. *On extension of Mittag-Leffler function*, Yokohama Math. J. 1968, vol. 16, no. 2 pp. 77–88.
60. H M Srivastava, *A certain family of sub-exponential series*, Int. J. Math. Educ. Sci. Tech, vol. 25, no. 2, 00211–216, (1994).
61. S.Westurlund, *Dead Matter has memory*. Physics Scripta 1991, vol. 43, pp. 174–179.
62. S.Westurlund et al *Capacitor theory* IEEE Trans on Dielectric and Electrical Insulation 1994, vol. 1, no. 5, pp. 826–839.
63. I. Podlubny *Presentation on fractional calculus* IEEE chapter control San Diego, September 29, 2005.
64. I. Podlubny, Ivo Petras et al, *Realization of fractional order controllers*, Acta Montanistica Slovak, Rocnik 8(2003) pp 233–235.
65. Fracncois Dubois Stephanie Mengue *Mixed Collocation for fractional differential equations*, Numerical Algorithms Dec. 2005 vol. 34, no. 2 pp 303–311.
66. Shequin Zhang, *Positive solutions for boundary value problems on non-linear fractional differential equations*. Electronic J. of differential equations 2006 no. 36 pp 1–12.
67. Cecilia Reis et al *Synthesis of logic circuit using fractional order dynamic fitness function*. Transaction on engineering computing and Technology ISSN ENFORMATICA VI 2004 1305-5313 pp 77–80.
68. J.D. Munkhammar *Fractional calculus and Taylor-Reimann series*. UUDM Thesis Report 2005.
69. V V Anh et al, *Fractional differential equation driven by Levy noise*. J. Appl. Math and stochastic analysis 16:2 (2003) pp 97–119
70. M Wallis *Computer experiments with fractional Gausian noises PART 3: Mathematical Appendix* Water Resource research 5, 1969, pp 328–338.

71. Vingare, Ivo Petras, Podlubny, et al, *Using fractional order adjustment rules and fractional order reference model in model reference adaptive control*, Non linear Dynamics 29, pp 269–279 (2002).
72. Kostadin Trenceveski et al. *On fractional derivative of some function of exponential type*. Ser Mat. University Beograd PUBL. ELEKTROTEHN FAK. Ser. Mat. 13(2002) pp 77–84.
73. Manabe. *Early development of fractional order control*. ASME DETC2003/VIB-48370 pp 1–8.
74. Xiaorang Li et al. *On concept of local fractional differentiation*. Department of applied mathematics University of Western Ontario London 2004 Lecture Notes.
75. Wang Jifeng, Yunkai et al. *Frequency domain analysis and application for fractional order control system* J of Phy. Conference Series 13(2005) pp 268–273.
76. C Yang et al. *A conceptually effective predictor corrector method for simulating fractional order dynamic control system*. ANZIAM. J. 47(EMAC2005) pp C168–C184.
77. Rafael Barcena et al *Auto tuning of fractional order hold circuit* IEEE Int. Conf on Control applications Sept 5–7, 2001, Mexico City pp 7–12.
78. I Podlubny. *Geometric and Physical meaning of fractional derivative*. Technical Uviversity Kosice Slovak.vol. 5 No. 4 (2002) ISSN 1311–045 Fractional Calculus and Applied Analysis 5(4) pp 367–386.
79. Mauro Bologna. *Short Introduction to fractional calculus*. Univ. Tarpeca Chile 2005. Lecture Notes pp 41–54.
80. R. K. Saxena *Fractional calculus and fractional differential equations* Lecture Note (chapter-3) pp 1–39, Jai Narayan Vyas University Rajasthan 2006.
81. R. K. Saxena, A. M. Mathai, H J Haubold, *On Generalized Fractional Kinetic Equations* Int J of Math. Phys. 0406046 v1 22-06-2004.
82. Juan Bisquert, Albert Compte, *Theory of electrochemical impedance of anomalous diffusion*, J of Electro analytical Chemistry 499(2001) pp 112–120.
83. A. Erdelyi *On some functional transformation* Univ Potitec Torino 1950.
84. Karabi Biswas, Siddharta Sen, Pranab Kumar Dutta, *Modeling of a capacitive probe in a polarizable medium*, Sensors actuators: Phys. 120 (2005) pp 115–122.
85. K. Biswas, S Sen, P K Dutta, *A constant phase element sensor for monitoring microbial growth*, Sensors and Actuators SNB9063 (2005) pp 1–6.
86. Shantanu Das, B B Biswas, *Fractional divergence for neutron flux profile in nuclear reactors*, International Journal Nuclear Energy Science and Technology. Vol.3, No.2 (2007) pp 139–159.
87. Shantanu Das, *Efficient control of Nuclear Plants* IAEA-TM-Control & Instrumentation 2007.
88. Shantanu Das, B. B. Biswas, *Total Energy Utilization from Nuclear Source*, PORT-2006, Nuclear Energy for New Europe Slovenia 2006.
89. Shantanu Das, et al. *Ratio control with logarithmic logics in new P&P control algorithm for a true fuel-efficient reactor.* Int. J. Nuclear Energy Science & Technology. Vol. 3, No. 1, 2007 pp. 1–18.
90. Ivo Petras, Vinagre, *Practical application of digital fractional order controller to temperature control*, Acta Montansitica Slovak 2002.
91. Chen, Vinagre, *A new type of IIR digital fractional order differentiator*, Signal Processing. 83(2003) pp 2359–2365
92. Chyi Hwang and Cheng, *A numerical algorithm for stability testing of fractional delay systems,* Elsevier Automatca 42(2006) pp 825–831.
93. Jack Leszcynski et al *A numerical method solution for ordinary differential equation of fractional order*. Technical university of Czestochowa Poland Lecture Notes 2005.
94. Kuo Diethelm and Ford, *Analysis of fractional differential equations* University of Manchester. Numerical Analysis Report No 377 pp 1–18 (2003).
95. K Diethelm, *An algorithm for the numerical solution of differential equations of fractional order*, Electronics transactions on Numerical Analysis, ISSN 1068-9613, vol. 5, March 1997, pp. 1–6.

96. Barcena et al *Discrete control for computer hard disc by fractional order hold devise.* University of Pais Vasc Spain. 2000 (pp 1–34)
97. Shih-Tong, H M Srivastava et al *A certain family of fractional differential equations,* Taiwanese Journal of Mathematics Sept-2000, vol. 4, no. 3, pp 417–426.
98. Chen, Vinagre, Podlubny et al *On fractional order disturbance observer* DETC03-VIB48371 ASME Design Engineering Technical Conference Chicago USA Sept 2-6, 2003. pp. 1–8.
99. Chen, Vinagre, Podlubny, *Fractional order disturbance observer for robust vibration suspension,* Non Linear Dynamics 38:pp. 355–367 (2004).
100. Lubomir Dorcak *Numerical Models for simulation of fractional order control system* Slovak Accedemy of Science 1994. UER04-94, pp 1–15.
101. Vinagre, Ivo Petras, *Two direct Tustin discretization methods for fractional order differentiation / integration.* J.Franklin Institute: 340 (2003) pp 349–362.
102. Lubomir Dorcak, Ivo Petras et al *Comparision of the methods for discrete approximations of fractional order operator* Acta Montanisca Slovak Rocnik 8(2003) cislo-4, pp 236–239.
103. Tom Hartley and Carl Lorenzo *Fractional order system identification by continuous order distribution* Elsevier Signal Processing 2003. pp 2287–2300.
104. S G Samko, A A Kilbas, O I Maritcheva *On Fractional Integrals and derivatives: Theory and Applications,* Gordon and Breach New York (1993).
105. Saha Ray S, *Exact solution for time fractional diffusion wave equation by decomposition method.* Physics Scripta (2007) pp 53–61.
106. Saha Ray S, Bera R K, *An approximate solution of non-linear fractional differential equation by Adomian's Decomposition,* Analytical Applied Math Computer 167 pp 561–71.
107. Saha Ray S Bera R K, *Analytical solution of Bagley Torvik equation by Adomian's decomposition method,* Appl. Math. Comp. 168 pp 389–410.
108. Agarwal O P, *Solution for fractional diffusion wave equation defined in bounded domain,* Non-Linear Dynamics 29 pp 145–155 (2002).
109. Ramiro S, Barbosa et al, *Time domain design of fractional differintegrators using least square,* Signal Processing 86(2006) pp. 2567–2581.
110. M.P. Lazarevie, *Finite Time Stability Analysis of PD^α fractional control of robotic time delays systems,* Mechanics Research Communications 33(2006) pp. 269–279.
111. Lynch V E, et al, *Numerical methods of solution of partial differential equations of fractional order,* J of Comput. Phy. 192(2003) pp 406–421.
112. Sierociuk Dominik, *Fractional Kalman Filter algorithm for states, parameters and order of fractional system estimation,* XV Krajowa Konferencja Automatyki, Warszawa, Polska (2005) pp 1–14.
113. Al-Alouni M A, *A class of second order integrators and low pass differentiators,* IEEE Trans. on circuit and systems 1: Fundamental Theory and Applications, 42(4) pp. 220–223 (1995)
114. Al-Alouni M A, *Filling the gap between the bilinear and backward difference transforms: an interactive design approach,* Int. J of Electrical Engineering Education, 34(4) pp. 331–337 (1997).
115. F.Riew, *Mechanics with fractional derivatives,* Phys. Rev. E, vol. 55, no. 3, pp 3581–3592, (1997).
116. H.E.Roman P A Alemany, *Continuous time random walks and fractional diffusion equation,* J Phys A: Math. Gen, vol. 27, pp. 3407–3410 (1994).
117. Scott Blair G W, *Psychoreology: links between the past and the present,* Journal of Texture Studies, vol. 5, pp. 3–12, (1974).
118. G W Scott Blair, *Measurement of Mind and Matter,* Dennis Dobson London 1950.
119. W.Wyss, *The fractional diffusion equation,* J. Math. Phys, vol. 27, no. 11, pp. 2782–2785 (1986).
120. W G Glockle, T F Nonnenmacher, *Fractional integrators and Fox functions in the thory of viscoelelasticity,* Macromolecules, vol. 24, pp. 6426–6436, (1991).
121. M Axtell, M E Bise, *Fractional calculus applications in control systems,* Proc of IEEE Nat. Aerospace and Electronics Conf, New York 1990, pp. 563–566.

122. H.H. Sun A.A. Abdelwahab B. Onaral, *Linear Approximation of Transfer function with a pole of fractional power.* IEEE Trans on Automatic Control vol. AC-29, no. 5, pp. 441–444, May 1984.
123. A. Charef, H H. Sun, Y.Y. Tsao, B. Onaral, *Fractal systems as represented by singularity function*, IEEE Trans. on Automatic Control, vol. 37, no. 9, pp. 1465–1470, Sept 1992.
124. H.E. Jones, B.A. Shenoi, *Maximally flat lumped element approximation to fractional operator immitance function*, IEEE Trans. on Circuits & Systems, vol. 17, no. pp. 125–128, 1970.
125. M. Moshrefi-Torbati & J.K. Hammond, *Physical and geometrical interpretation of fractional operators*, Journal of Franklin Institute, vol. 335, no. 6, pp. 1077–1086, 1998.
126. M. Nakagawa & K. Sorimachi, *Basic characteristics of fractance device*, IEICE Trans fundamentals, vol. E-75-A, no. 12, pp. 1814–1818.
127. M. Ichise, Y. Nagayanagi, T. Kojima, *An analog simulation of non-integer order transfer function analysis for electrode process*, J. Electroanal. Chem. Vol. 33, no. 1, pp. 253–265, 1971.
128. Bohennan Gary et al. *Electrical components with fractional order impedances*, Patent US 20060267595, November 2006.
129. V.Kirayakova, *Generalized Fractional Calculus and Applications*, Pitman Research Notes in Maths. No. 301, Longman, Harlow, 1994.
130. F. Mainardi, *Fractional relaxation-oscillation and fractional diffusion wave phenomena*, Chaos Solitons and Fractals, vol. 7, 1996, pp. 1467–1477.
131. R. Gorenflo, F. Mainardi, *Fractional calculus: integral and differential equations of fractional order*, Fractals and Fractional Calculus in Continuum Mechanics Springer-Verlag, Vienna-New York, 1997.
132. R. Gorenflo, Yu. Luchko and S. Rogosin, *Mittag-Leffler Type Functions: Notes on Growth Properties and Distribution of Zeros*, no. A-97-04. Department of Mathematics and Informatics, Free University of Berlin, 1997.
133. R. Gorenflo, *Fractional Calculus: Some numerical methods*, Fractals and Fractional Calculus in Continuum Mechanics Springer-Verlag, Vienna-New York, 1997.
134. M.Caputo and F. Mainardi, *A new dissipation model based on memory mechanism*, Pure and applied Geophysics, vol. 91, no. 8, 1971, pp. 134–147.
135. M.Caputo, *The rheology of an anelastic medium studied by means of the observation of splitting of its eigenfrequencies*, J. Acoust. Soc. Am.vol. 86, no. 5. 1989 pp. 1984–1987.
136. Bush. V. *Operational Circuit Analysis* Wiley New-York (1929)
137. LePage. W R *Complex variables and the Laplace transform for Engineers* Dover, New-York (1961).
138. Oberhettinger.F and Badii L, *Tables of Laplace transforms*, Springer-Verlag Berlin (1973).
139. Robotnov Y N, *Elements of Hereditary Solid Mechanics*, English MIR Publishers Moscow (1980).
140. Heaviside. O. (1922) *Electromagnetic Theory,* Chelsea Edition New-York (1971).
141. Glockle W G and Nonnenmacher T F, *Fractional Integral Operators and Fox Functions in the Theory of Viscoelasticity*, Macromolecules, vol. 24, pp. 6426–6434 (1991)
142. V Churchill Ruel, W Brown James and F Verhey Roger, *Complex Variables & Applications*, Third Edition Mc Graw Hill Kogakusha Ltd. (1974).
143. M.E.Rayes-Melo, J.J.Martinez-Vega, C.A.Guerrero-Salazar, U.Ortiz-Mendez, *Modeling of relaxation phenomena in organic dielectric materials, applications of differential and integral operator of fractional order*, Journal of Optoelectronics and Advanced materials, Vol. 6, No. 3, September, 2004 pp1037–1043.
144. Marco Raberto, Enrico Scalas, Francesco Mainardi, *Waiting-times and returns in high-frequency financial data: an empirical study.* Elsevier ARXIV: COND-MAT/0203596V1 28 (2006) pp1-8
145. Enrico Scalas, Rudolf Gorenflo, Francesco Mainardi, Marco Raberto, *Revisiting the derivation of the fractional diffusion equation*, Elsevier ARXIV: COND-MAT/0210166 V2 11.10.2002 (2002).

146. Rina Schumer, David.A.Benson, Mark.M.Merschaert, Stephen.W.Wheatcraft, *Eulerian derivation of fractional advection-dispersion equation,* J. of contaminant hydrology 48(2001) Elsevier pp69–88.
147. Chien-Cheng Tseng, Soo-Chang-Pei, Shih-Cheng Hsia, *Computation of fractional derivatives using Fourier transforms and digital FIR differentiator,* Elsevier Signal Processing 80(2000) pp151–159.
148. Andrea Roceo, Bruce. J. West, *Fractional calculus and evolution of fractal phenomena,* Elsevier Physica A 265(1999) pp535–546.
149. R. Hifler, *Fractional Diffusion based on Riemann-Liouvelli Fractional Derivatives,* J. Phys. Chem B Vol. 104 (2000) p3914.
150. H.M.Srivastava, S. Owa, *Some characterization and distortion theorems involving fractional calculus, generalized hyper geometric functions, Hadamard products, Linear Operators and certain subclasses of Analytical functions,* Nagoya Math. J. Vol.106 (1987) pp1-28.
151. Rina Schumer, David .A. Benson, *Multiscaling fractional advection dispersion equation and their solution,* Water Resource Research Vol.39, No.1, pp12-1-12-11. (2003)
152. R. Gorenflo, F. Mainardi, *Fractional oscillators and Mittag-Leffler function,* Technical Report Preprint A-14/96, University of Berlin, (1996).
153. H. Bateman, *Higher transcendental function,* Vol. 3, Mc Graw Hill New-York (1954).
154. A. Erdelyi, *Asymptotic Expansions,* Dover (1954).
155. R. Hifler, *Applications of fractional calculus in Physics,* World Scientific Singapore (2000).
156. R. Hifler, L. Anton, *Fractional master equation and fractal time random walks,* Phys Rev. E51 (1995) R818–R851.
157. J. Klafter, A. Blumen, M. F. Shlesinger *Stochastic pathways to anomalous diffusion,* Phys Rev A 35 (1987).
158. I.M. Gel'fand, G.E Shilov, *Generalized functions,* Vol.1 Academic Press New-York (1964).

This is the beginning

Printing: Krips bv, Meppel, The Netherlands
Binding: Stürtz, Würzburg, Germany